Methuen's Monographs on Biological Subjects
General Editor: Michael Abercrombie

RESPIRATION IN PLANTS

RESPIRATION IN PLANTS

by

WALTER STILES
M.A., Sc.D., F.L.S., F.R.S.
EMERITUS PROFESSOR OF BOTANY IN THE UNIVERSITY OF BIRMINGHAM

and

WILLIAM LEACH
D.Sc.
PROFESSOR OF BOTANY IN THE UNIVERSITY OF MANITOBA

WITH TEN DIAGRAMS

LONDON: METHUEN & CO. LTD.
NEW YORK: JOHN WILEY & SONS, INC.

First Published April 29th 1932
Second Edition, Revised, July 1936
Third Edition, Revised, 1952

3.1
CATALOGUE NO. 4113/U

PRINTED IN GREAT BRITAIN

PREFACE

THE supreme importance of respiration, being as it is one of the most universal and fundamental processes of living protoplasm, is recognized by all physiologists. In spite of this, students of botany frequently give respiration little more than a passing consideration.

This curious state of affairs is largely due to the fact that most of the existing accounts of respiration in plants are unsatisfactory because they are either insufficiently comprehensive or insufficiently lucid. In the present book we have aimed at giving an account of the nature of plant respiration which is readable and understandable by the elementary student of botany and which at the same time contains sufficient information to render it of value to the advanced student. We have throughout endeavoured to indicate the principles of plant respiration rather than to catalogue a mass of detailed observations from researches often of very dubious value. Nor have we thought it desirable to enter into a detailed discussion of the very considerable amount of recent work dealing with oxidizing systems in yeast and in animal cells, and the theories based on this work, for the bearing of these observations on the problems of plant respiration is at present not in the least clear.

Reference has, however, been made from time to time to actual researches where we have considered them to be useful as illustrations of various aspects of the subject. A list of these works is given at the end of the book. No attempt has been made to compile a comprehensive bibliography of books and papers dealing with plant respiration.

W. S.
W. L.

BIRMINGHAM, *1932*

PREFACE TO THE THIRD EDITION

DURING the fifteen years that have elapsed since the appearance of the second edition of this book very considerable advances have been made in our knowledge of the enzyme systems concerned in the breakdown of carbohydrate in plant tissues. Also, recent work has shown that a knowledge of the oxidizing systems in some animal tissues may be of assistance in elucidating problems of plant respiration. In view of the changed outlook that these developments have produced, the last chapter has been largely rewritten. Advances in knowledge of other aspects of respiration in plants have also necessitated a number of important alterations and additions throughout the book.

W. S.
W. L.

BIRMINGHAM, ENGLAND
WINNIPEG, CANADA
 February 1951

CONTENTS

CHAP.		PAGE
I	INTRODUCTORY	1
II	RESPIRATION OF NORMAL PLANTS UNDER AEROBIC CONDITIONS	13
III	ANAEROBIC RESPIRATION	59
IV	THE MECHANISM OF RESPIRATION	81
	LITERATURE CITED IN THE TEXT	143
	INDEX	161

CHAPTER I

INTRODUCTORY

IN the maintenance of its life, every living thing exhibits a phenomenon which consists essentially in the breaking down of complex substances into simpler ones, with consequent release of energy. This phenomenon has been called 'dissimilation' in contrast to 'assimilation' in which simple substances are absorbed and built up into the organism in the form of substances of greater complexity and higher energy content. Although this dissimilation affects different materials in different species, it very commonly involves the breaking down, by oxidation, of carbohydrates and fats, the end-products being carbon dioxide and water. The dissimilation process thus involves an exchange of gases between the organism and its environment, oxygen being absorbed and carbon dioxide evolved. This exchange of gases, so characteristic in animals, is equally characteristic of the vast majority of plants. Hence the term 'respiration', used to denote this gaseous exchange and the processes of which it forms a part, is equally applicable to animals and plants.

Although the term respiration at first referred to the exchange of gases between the organism and its environment, so that, in the case of animals, it was synonymous with the term breathing, it has for many years now been more usual to regard respiration as involving the whole of the dissimilation process. The leading workers of the second half of the nineteenth century, such as Sachs, Pfeffer and Palladin, who were responsible for the modern conception of respiration, all gave the word respiration this wider meaning, and in this book respiration in plants is taken to include all the phenomena of dissimilation, the characteristics of which are the breaking down of complex substances into simpler ones with a consequent release of energy.

Respiration, so defined, is a much more fundamental property of living substance than the exchange of gases between organism and environment, for gaseous exchange is merely an aspect of the most usual form of respiration, and may not always be present, whereas respiration is a property, not merely of every living organism, but of every actively living cell. At the same time the cases in which respiration does not involve an exchange of gases are relatively few, and it is no wonder that Sachs, in reviewing the characteristics of plant respiration, laid particular stress on the importance of a supply of oxygen.

Indeed, lately some writers would limit the term respiration to processes in which substances are broken down through oxidation by molecular oxygen. There are other dissimilation processes included in the conception of respiration indicated above which do not involve such an oxidation. These processes, which have for many years been regarded as coming within the category of respiration processes, would be excluded from this modern view of respiration. There are arguments for and against such an exclusion, but whether or not such processes are so excluded no account of respiration in plants would be adequate without a consideration of them. Without any intention of dogmatizing on a matter which is largely, though not entirely, a matter of definition, we shall in this book use the word respiration in its wider sense.

We may regard the history of our knowledge of respiration in plants as beginning in the seventeenth century with such observations as that of Malpighi, published in the year 1679, that seeds require a supply of air in order to germinate. It was, however, naturally not until the development of pneumatic chemistry by Priestley, Lavoisier, and others, that the nature of gaseous exchange between organisms and environment could be appreciated. By 1777 Scheele had shown that germinating seeds absorbed and utilized oxygen and produced carbon dioxide, while about the same time Lavoisier began his work on animal respiration which was to put knowledge of that subject on a sound basis. In 1779 Ingen-Housz, in his *Experiments upon Vegetables*, showed that all living plants give out carbon dioxide in the dark, and that non-green plants do so in the light as well.

The serious study of plant physiology began with the introduction of quantitative investigation by de Saussure. In a paper published in 1797 with the title *La formation de l'acide carbonique est-elle essentielle à la végétation?*, he laid emphasis on the similarity between plants and animals in their production of carbon dioxide, and in their absorption of oxygen from the atmosphere for the formation of carbon dioxide. By actual measurement he was able to show that the volume of oxygen absorbed by germinating seeds was equal to that of the carbon dioxide produced. He dealt at greater length with the subject in his *Recherches chimiques sur la végétation*, published in 1804. In this work he recorded that with leaves in the dark he found that less carbon dioxide was evolved than oxygen absorbed. He further showed that different leaves respired at very different rates, while he observed the gaseous exchange exhibited by respiring roots, flowers, and fruits.

De Saussure clearly distinguished between the assimilatory gaseous exchange which proceeds in the green parts of plants in the light and the reverse gaseous exchange which proceeds in non-green plants in both light and darkness and in green plants also in the dark. He further showed that germination and growth are dependent on a supply of oxygen.

In a later work, published in 1822, de Saussure showed that the evolution of heat by flowers, which had been recorded by Lamarck in *Arum italicum* in 1778, was accompanied by absorption of oxygen, two phenomena which are both features of respiration.

During the next forty years little progress in knowledge of respiration was made. Throughout this period much confusion of thought appears to have resulted through both gaseous exchanges being called 'respiration'. A plant was said to exhibit a diurnal respiration during the day and a nocturnal respiration at night. Nor is it likely that progress in knowledge of these matters was much helped when such an authority as Liebig denied the existence of a respiration in plants comparable with that of animals. According to him, plants simply absorbed carbon dioxide from the air or soil and later gave it off unchanged when assimilation stopped, much in the same way as water vapour is given off in transpiration.

However, von Mohl, in his *Grundzüge der Anatomie und Physiologie der vegetabilischen Zelle* published in 1851, and also Garreau in the same year, made perfectly clear the difference between these two kinds of gaseous exchange, and definitely indicated the significance of both in the life of the plant. It was not, however, until 1865 that Sachs pointed out what he later called 'the scarcely conceivable thoughtlessness and obtuseness' in 'speaking of a double respiration of plants—of a so-called diurnal respiration, meaning assimilation, and a so-called nocturnal respiration, by which was understood the evolution of carbon dioxide which occurs in true respiration'. From this time onwards the term respiration ceased to be used in connexion with the assimilatory process.

The work and writings of Sachs gave a great impetus to plant physiology, and, from his time, work on respiration has proceeded practically without interruption. Among the numerous contributors to our knowledge of the subject during the last seventy years, perhaps special mention should be made of Pfeffer and Palladin as original workers themselves and as inspirers of many others. During this period research on respiration has chiefly aimed at acquiring information regarding the magnitude of the process and the manner in which it is affected by external and internal conditions. This has been largely with a view to discovering what is generally called the mechanism of the process, such questions being involved as the nature of the materials utilized, the stages in the process, its relation to cell enzymes, the way in which it is linked with other plant processes and growth, and the part it plays in the life of the plant generally.

At the beginning of this chapter it was pointed out that the essential characteristic of what is now called respiration is not a particular exchange of gases between organism and environment, although this is generally present, but a catabolism or breaking down of more complex substances into simpler ones with a release of energy. In the commonest form of respiration this release of energy is brought about by the oxidation of organic material such as carbohydrates, fats, and proteins, for which a supply of atmospheric oxygen is necessary. This process is known as aerobic or oxygen respiration, and is universal enough to be regarded as the normal mode of respiration in plants. Indeed, as already indicated, some recent writers would limit the term respiration, in higher plants at any rate, to an oxidative breakdown of this kind. There will obviously be differences in detail according as the material utilized, or *substrate*, is carbohydrate, fat, or some other substance.

There are, however, other processes met with in plants which bring about a release of energy. The most important of these is that which has for many years been known as anaerobic respiration in which carbohydrate is broken down to alcohol and carbon dioxide without the participation of atmospheric oxygen, and which is thus similar to, and possibly identical with, the process known as alcoholic fermentation brought about by yeast. This process, as we have seen, would not be included as a respiratory process by some recent writers, who speak of the process as fermentation. This assumes that the process is indeed identical with alcoholic fermentation, but as the amount of alcohol produced in the anaerobic breakdown of the substrate in higher plants appears frequently to be less, and in some instances very much less, than the amount produced in alcoholic fermentation, the substitution of fermentation for anaerobic respiration is not free from objection. Meirion Thomas has used the term zymasis to describe the processes in which carbohydrate is broken down in the plant to yield carbon dioxide and alcohol. Probably all plants which normally respire aerobically continue to respire anaerobically, for a time at any rate, when deprived of oxygen. Much work has been done with the object of determining the relationship between these two kinds of respiration.

Anaerobic respiration is normally met with in certain bacteria. Some of these live only in absence of oxygen or in presence of a negligible concentration of this gas. Such organisms are termed *obligate* anaerobes, and include among others such forms as *Bacillus denitrificans* and certain butyric and lactic bacteria. *Facultative* anaerobes, on the other hand, are organisms which normally require oxygen but which can live anaerobically when grown on suitable media. Certain butyric and lactic bacteria also fall into this class as well as *Bacillus phosphorescens* and various thermophile bacteria. Among bacteria there also occur oxidations which appear to serve a respiratory function, and in which not an organic substrate, but an inorganic one, is oxidized. The best known of these are the nitrifying bacteria, *Nitrosomonas* and *Nitrococcus*, which obtain energy by oxidizing the ammonia of ammonium salts to nitrites. The oxidation is usually represented by the equation:

$$2NH_3 + 3O_2 = 2HNO_2 + 2H_2O$$

A second type of oxidation is present in the nitrating bacteria *Nitrobacter* which oxidize nitrites to nitrates:

$$2HNO_2 + O_2 = 2HNO_3$$

Apparently the energy obtained by these respiratory processes is sufficient for the life of these organisms, for there appears to be no respiration of organic material in them. Further, with the energy so obtained they are able to assimilate carbon dioxide without the necessity of absorbing light energy as in green plants.

Similar to the nitrifying and nitrating bacteria are the sulphur bacteria, including *Beggiatoa*, *Thiothrix*, and *Hillhousia*, which utilize hydrogen sulphide for respiratory purposes. The hydrogen sulphide is oxidized to sulphuric acid, free sulphur being formed in an intermediate stage and appearing in the cell in the form of relatively large particles:

$$2H_2S + O_2 = S_2 + 2H_2O$$
$$S_2 + 3O_2 = 2SO_3$$

A further group of bacteria, the thiosulphate bacteria, *Thiobacillus*, oxidize thiosulphates to sulphates:

$$6K_2S_2O_3 + 5O_2 = 4K_2SO_4 + 2K_2S_4O_6$$

The iron bacteria, *Spirophyllum ferrugineum*, *Crenothrix polyspora*, and others, are said to utilize ferrous iron for respiration, oxidizing it to the ferric condition. It has been suggested that the action is as represented in the following equation:

$$2Fe(HCO_3)_2 + OH_2 + O = Fe_2(OH)_64 + CO_2$$

Some doubt has, however, been cast on the view that these bacteria utilize ferrous salts in this way.

The hydrogen bacteria, *Hydrogenomonas spp.*, *Bacillus hydrogenes*, and *B. pantotrophus*, oxidize hydrogen to water, and, like the other forms mentioned above, obtain enough energy from this reaction to enable them to assimilate carbon dioxide without a supply of radiant energy.

These various kinds of respiration met with in bacteria are interesting and important in that they help to indicate the meaning of the respiratory process. They are, all the same, limited to a very few organisms, which, although plants, belong to a very highly specialized group. They will not be dealt with further in this book.

It has been noted that the essential property of

respiration, whatever form it may take, is the release of energy. Every actively living cell respires and therefore presumably requires a supply of energy. This is generally accepted as a fact, but the reasons for it are often stated in the vaguest terms. One considerable worker in this field states that 'vital combustion . . . causes the mysterious apparatus of the living protoplasm to function', another that the 'energy is used in other processes that go on within the plant'. There are indications that part, perhaps much, of the energy released in respiration is dissipated in the form of heat, and can be measured as such, and that only a small proportion is transformed into mechanical or chemical energy. Some is presumably used in the building up of more complex compounds from inorganic materials and the sugars formed in photosynthesis, and in the building up of protoplasm itself, but it is not known, as Pfeffer pointed out fifty-three years ago, whether or not the respiratory processes involve a continual destruction and re-formation of the protoplasm. Some energy is no doubt utilized in streaming movements of protoplasm and other movements of material in the plant. The passage of salts into and through the plant would also appear to require energy provided by respiration. But, apart from the few plant bodies that have the power of active locomotion, there does not appear anywhere in the vegetable kingdom a necessity for a large supply of energy for the mere maintenance of life comparable with that obviously required by a free-moving animal. Nevertheless, we must suppose that a certain and continuous supply of energy is as necessary for the plant as for the animal, for of all the characteristics of living matter, none is more constant than the presence of respiration, and nothing is more characteristic of the actively living plant cell than the continuous incidence of this process.

It has been stated above that de Saussure initiated a new era in plant physiology in that he introduced quantitative measurement into his researches. The advance of knowledge of respiration, as of every plant process, has depended largely on its measurement. We have noted that the respiratory process is not constant throughout the plant kingdom and that its essential characteristic, a release of energy, may be effected in different ways. However, in the vast majority of cases respiration consists of a slow oxidation of material, of which the outward signs are a consumption of oxygen and elimination of carbon dioxide. This is the process which, as we have already indicated, is known as aerobic, or oxygen respiration, or sometimes as normal respiration. Here, theoretically, respiration could be studied quantitatively by determining either the oxygen consumption or the carbon dioxide evolution exhibited by the respiring tissue, and in practice the determination of one or other of these quantities usually forms the basis of respiration measurement. The loss of material in respiration would also give a measure of the process, but such determinations are not always practicable. The measurement of respiratory activity, therefore, generally resolves itself into either a determination of oxygen absorption or carbon dioxide evolution.

Where the substrate of respiration is a carbohydrate the complete oxidation of the carbohydrate to carbon dioxide and water involves the consumption of a volume of oxygen equal to that of the carbon dioxide evolved according to the general equation:

$$C_mH_{2n}O_n + mO_2 = mCO_2 + nH_2O$$

When this relation is actually maintained it does not matter whether the respiration is measured by determining oxygen absorption or carbon dioxide evolution. Not infrequently, however, as will be seen in the next chapter, the volumes of oxygen absorbed and carbon dioxide evolved, owing to a number of reasons, are not the same. For example, a plant which normally respires aerobically will still give out carbon dioxide in absence of oxygen, and in low concentrations of oxygen the volume of carbon dioxide evolved may exceed that of oxygen absorbed, as if respiration were partly aerobic and partly anaerobic. In such cases, and wherever there is evidence of a change in the ratio of carbon dioxide evolved to oxygen absorbed, as well as in many other instances, a knowledge of both the rate of oxygen absorption and carbon dioxide evolution is desirable.

Various forms of apparatus have been devised for measuring the oxygen absorbed and the carbon dioxide evolved by respiring plants. In the simplest types of apparatus the respiring material is enclosed in a vessel containing a gas mixture of known composition. After a lapse of a suitable time the carbon dioxide is determined by observing the reduction in volume at constant pressure after this gas is absorbed by potassium hydroxide, while the oxygen can be similarly determined by the use of pyrogallol. Where the volume of gas available is sufficient for exact determination, the changes in composition of the gas mixture can be measured by direct analysis. Frequently, however, the quantity of gas available is insufficient for this, and in consequence of the need for determining the carbon dioxide and oxygen changes in relatively small quantities of gas, various so-called micro-eudiometers have been devised, one of the earliest of which was that of Bonnier and Mangin. The principle involved in the use of this apparatus was, however, the same as that of an ordinary gas-analysis method. It was claimed that oxygen could be determined to 0·5 per cent. and carbon dioxide to 0·3 per cent. of the total volume.

A further variant of this method consists in absorbing the carbon dioxide evolved in a solution of an indicator. The colour change produced depends upon the amount of carbon dioxide absorbed and can be estimated by matching the indicator with standard tints.

Of late years manometric methods for measuring respiration have become popular. The respiring material is held in a vessel containing an absorbent of carbon dioxide, usually potassium hydroxide or sodium hydroxide, and to which a manometer is attached. As respiration proceeds oxygen is absorbed by the tissue and the carbon dioxide evolved is removed by the absorbent, so that there is a loss of gas in the vessel to the extent of the oxygen absorbed and a consequent movement of the liquid in the manometer which thus provides a measure of the respiration in terms of oxygen consumed. A number of convenient forms of manometer have been devised, with which the names of Barcroft, Warburg, Thunberg and Fenn are associated. With the use of small respiration vessels the method can be made quite sensitive. Also, by the carrying out of parallel experiments with exactly similar samples of material, if such are possible, in one of which the carbon dioxide absorbent is present in the respiration vessel and in the other of which it is absent, values for the evolution of carbon dioxide as well as of the absorption of oxygen can be calculated. A single value for the evolution of carbon dioxide can also be obtained if, in the vessel containing the respiring tissue and the carbon dioxide absorbent, acid is added to the absorbent at the end of the experiment so that the absorbed carbon dioxide is released.

The authors have developed the instrument known as the katharometer for the measurement of carbon dioxide evolution. Here the change in resistance of a spiral of platinum wire consequent on changes in the concentration of carbon dioxide in the gas surrounding the wire forms the basis of the measurement. By taking necessary precautions, carbon dioxide can be determined to 0·001 per cent. of the total volume. This instrument is thus 300 times as sensitive as the apparatus of Bonnier and Mangin and can be used for very small quantities of gas. It has also an advantage over most other methods of measuring respiration in that a continuous record of carbon dioxide evolution can be obtained with it.

It is frequently an advantage in measuring respiration not to keep the respiring tissue in a closed chamber but to pass a continuous current of gas of known composition over the material. The carbon dioxide is then absorbed from the gas after it leaves the respiration vessel. In the most usual form of apparatus the gas bubbles through a tube, the well-known Pettenkofer tube, containing a standard solution of barium hydroxide, for a definite time, the carbon dioxide absorbed being then determined by titration. Instead of determining the carbon dioxide by titration, Spoehr measured the electrical conductivity of the solution, the fall in electrical conductivity being a measure of the carbon dioxide absorbed. Other workers have absorbed the carbon dioxide in potassium hydroxide solution and determined the quantity of gas absorbed either from the gain in weight or by titration.

The chief disadvantage of using a continuous stream of gas is that the method is not very sensitive, so that it frequently requires a considerable amount of material and a long period of respiration in order to obtain a single measurement. The chief advantages of the method are that the carbon dioxide does not accumulate in the neighbourhood of the respiring tissue and that a series of measurements can be made over a period of time.

However, both these advantages can be introduced into methods involving the use of a closed system if the respiration chamber forms part of a circulatory system involving also a vessel containing the absorber of carbon dioxide and a pump to effect circulation.

The rate of carbon dioxide evolution or of oxygen absorption having been measured, there still remains the question of how these values can be used to express respiratory activity. Usually the carbon dioxide evolved in unit time per unit of dry matter is taken as a measure of respiratory activity. Palladin attempted to calculate respiratory activity in terms of evolution of carbon dioxide per unit of protoplasm, but it is doubtful if he was really able to obtain a value for the amount of protoplasm in different tissues.

We shall have occasion to refer to this question of measurement of respiratory activity in the next chapter.

CHAPTER II

RESPIRATION OF NORMAL PLANTS UNDER AEROBIC CONDITIONS

MOST of the common plants with which we are familiar live with the greater part of their external surfaces in contact with the atmosphere. There is therefore available for their use an abundant supply of oxygen. It has already been stated that plants in their normal respiration absorb quantities of oxygen and at the same time give out carbon dioxide. This exchange of gases, which is the outward manifestation of respiration, although continually taking place, may be masked or even reversed in the green parts of the plants when they are exposed to light, as a result of photosynthetic activity. During night time, or when a plant is placed in the dark, the absorption of oxygen and evolution of carbon dioxide can always be demonstrated. The exchange of gases takes place over all parts of the plant which are in contact with the atmosphere except where the external walls of the superficial cells are rendered impermeable by impregnation with such substances as cutin and suberin. Within the plant the process is maintained by diffusion between one living cell and another and between the cells and intercellular spaces, the latter usually being in direct communication with the outside air through such channels as stomata and lenticels.

The precise nature of the process which we call respiration is, as yet, little understood. Its existence and intensity can easily be demonstrated and studied through the medium of the resulting gaseous exchange which is nearly always taking place. We know that it resembles a slow combustion whereby certain complex organic substances are oxidized and broken down into simple substances with an accompanying release of energy. The greater part of this released energy is frequently dissipated in the form of heat, and as such can be detected and measured.

The amount of energy that is released when a complex substance is broken down into simple substances is equal to the amount of energy that has to be supplied to the simple substances in order to make them combine and form the complex substance. In the green parts of plants we know that energy from the sun in the form of light is absorbed and used in photosynthesis to bring about the formation of sugar from carbon dioxide and water. This reaction is indicated in a general way by the equation:

$$6CO_2 + 6H_2O + \text{energy (light)} = \underset{\text{(sugar)}}{C_6H_{12}O_6} + 6O_2$$

In the presence of atmospheric oxygen, sugar can readily be made to burn and give out heat, and in the process of burning it is broken down into carbon dioxide and water, as shown in the equation:

$$C_6H_{12}O_6 + 6O_2 = 6CO_2 + 6H_2O + \text{energy (heat)}$$

The above equation represents the process in its simplest form in which direct combination occurs between gaseous oxygen and sugar. The combination is rapid so that the whole of the energy involved is released as heat in a very short time, which consequently causes a marked rise in temperature in the neighbourhood of the seat of the reaction. In plant respiration we frequently find sugar being broken down to carbon dioxide and water, but the process is more complicated. It involves a chain of reactions in which a series of intermediate products is formed. Also the quantity of sugar oxidized is relatively small and is distributed through a relatively large mass of tissue. As a consequence, even where practically all the energy is known to be released in the form of heat, the resulting rise in the temperature of the tissue is so small as to be often difficult to measure. This chain of reactions, which is discussed at length in

Chapter IV and which results in the breaking down of sugar, is due to the activities of enzymes that are produced within the protoplasm of the living cells where respiration is taking place. It is thus misleading to speak of respiration as combustion. There are, however, two points of resemblance between respiration and combustion; firstly, the substrate and final products may be the same in the two processes, and secondly, the total amount of energy set free will be the same in the two processes provided the end products are the same.

RESPIRATORY QUOTIENT

It will be readily gathered from an examination of the equation on page 14 that where, in the plant organ, sugar is the substance broken down during respiration under conditions of a plentiful supply of oxygen, with the production of carbon dioxide and water, six molecules of oxygen will be used up for every molecule of hexose sugar respired. As a result of this, six molecules of carbon dioxide are set free. In other words, the ratio of the volume of carbon dioxide evolved to the volume of oxygen absorbed is equal to unity. This ratio is known as the *respiratory quotient*. There is thus an intimate relationship between the value of the respiratory quotient and the composition of the respiratory substrate on the one hand, and the nature of the respiratory process on the other. Where this substrate is carbohydrate the respiratory quotient is always in the neighbourhood of unity if the respiratory process results in complete breakdown to water and carbon dioxide. This has been demonstrated by a number of investigators, by measuring the oxygen intake and carbon dioxide output of respiring fungus mycelia growing on culture solutions containing various known substances.

For example, Puriewitsch obtained the values given below for the respiratory quotient in the case of *Aspergillus niger*.

TABLE I

RESPIRATORY QUOTIENTS OF *ASPERGILLUS* ON VARIOUS MEDIA
(*From* Puriewitsch)

10 per cent. Sucrose	10 per cent. Glucose	10 per cent. Raffinose
1·05	1·17	0·90
1·09	1·19	0·93

Similarly, De Boer obtained values for the respiratory quotient of between 0·99 and 1·21 for *Phycomyces* grown on bread.

In the case of higher plants it is not always easy to correlate respiratory quotient values with the composition of the substrate, as the latter may often be difficult to ascertain. With leaves, however, the problem is fairly simple, as these organs, if removed from the plant during active assimilation, contain abundance of carbohydrates. These carbohydrates, if the leaves are placed in the dark, form the substrate for respiration, and, just as with the fungi mentioned above, the respiratory quotient approaches unity. The table on page 17 gives values of the respiratory quotients in a number of leaves investigated by Maquenne and Demoussy.

It frequently happens in plants that fats and not carbohydrates form the substrate for respiration. Fats require a larger amount of oxygen for their complete oxidation to water and carbon dioxide than is the case with carbohydrates. Thus the complete oxidation of the fat tripalmitin, $C_3H_5O_3(OC.C_{15}H_{31})_3$, involves the utilization of 145 molecules of oxygen for every 102 molecules of carbon dioxide produced:

$$2C_{51}H_{98}O_6 + 145O_2 = 102CO_2 + 98H_2O$$

As a consequence, we find that respiration involving the breaking down of fats into carbon dioxide and water results in a respiratory quotient of less than unity. This fact also has been clearly demonstrated by means of the fungus *Phycomyces*. By growing *Phycomyces* on a ground-linseed medium, De Boer obtained values for the respiratory quotient varying between 0·66 and 0·75.

TABLE II

RESPIRATORY QUOTIENTS OF LEAVES

(*From* Maquenne and Demoussy)

Ailanthus	1·08	Pea	1·07
Aspidistra	0·97	Pear	1·10
Aucuba	1·11	Poppy	1·09
Begonia	1·11	Privet	1·03
Cherry Laurel	1·03	Rhubarb	1·02
Chrysanthemum	1·02	Ricinus	1·03
Haricot	1·11	Rose	1·02
,,	1·07	Spindle Tree	1·08
Ivy	1·08	Sorrel	1·04
Lilac	1·03	Tobacco	1·03
Lily	1·07	Turnip	1·11
Mahonia (autumn)	0·95	Vine	1·01
Maize	1·07	Wheat	1·03
Oleander	1·05	Wild Grape	1·00

The respiratory quotient has formed a centre of interest in a considerable number of researches on the course of respiration of seeds during germination. As is well known, seeds contain reserve stores of food materials which provide for the needs of the growing plant in its early stages of development. In the majority of seeds this food reserve consists mostly of oil (liquid fat); in some, for example, those of Leguminosae and Gramineae, it may be mainly carbohydrate in the form of starch, or less frequently, hemicellulose or sugar, while in others it appears to be largely protein.

From what has already been said it is clear that the nature of the food reserve in any given seed should affect the value of the respiratory quotient during germination. There is ample experimental evidence to show that this is so, and in general it may be stated that in seeds with carbohydrate reserve materials the respiratory quotient during the greater part of the germination period approaches unity. On the other hand, with fat-containing seeds, the respiratory quotient falls considerably below unity. The matter, however, is far from being simple, and although a number of published figures showing the changes which the respiratory quotient undergoes during germination of certain seeds are available, the divergences which exist between the results of one investigator and those of another in any given species leave much to be desired. In spite of the lack of reliable information as to the absolute values of the respiratory quotient of germinating seeds, the work of different authors shows some measure of agreement with regard to the way in which it varies during the germination period.

Apparently during the first few hours of germination of all the seeds examined, the respiratory quotient is in the neighbourhood of unity or is greater than unity. As germination proceeds the value of the quotient falls. In seeds with carbohydrate food reserve the fall in the value of the quotient may be slight, and in any case continues only for a short time, after which it rises again and approaches unity. With fat-containing seeds the fall in the value of the quotient is very marked and continues usually for a considerable time. Eventually, however, it also rises and approaches unity. This difference in the behaviour of the respiratory quotient for carbohydrate and fat-containing seeds is illustrated by the curves shown in Fig. 1.

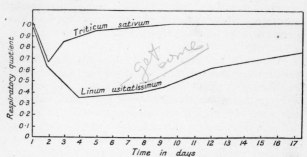

FIG. 1.—Curves showing the changes occurring in the value of the respiratory quotient during the first 17 days of germination and development of seeds of *Triticum sativum* and *Linum usitatissimum*

(*After* Bonnier and Mangin)

It thus appears that the germination of fat-containing seeds can be, broadly speaking, divided into three phases. *Phase* 1: an initial period during which the value of the respiratory quotient is high, or in other words, the output of carbon dioxide is approximately equal to or greater than the intake of oxygen. *Phase* 2: a middle period during which the respiratory quotient decreases to a minimum value owing to a considerable increase in the volume of oxygen absorbed compared with the volume of carbon dioxide given out. *Phase* 3: a final period in which the respiratory quotient rises and may approach unity. Theoretical explanations which would appear to account for the differences that exist between these three phases will now be briefly discussed. In the initial phase the respiratory quotient is in the neighbourhood of unity. This is probably due to the fact that when germination is beginning, the respiratory substrate is provided by a small quantity of hexose sugar which is always present in dormant seeds. As water continues to be absorbed by the seed the enzymes which bring about the conversion of fats into carbohydrates are activated with the result that the respiratory quotient falls. This fall in the quotient which is shown by both typically carbohydrate-storing seeds as well as fat-storing seeds is due to the fact that even in carbohydrate seeds there is present a small amount of fat.

We thus note that phase 2 with its falling quotient is the result of the utilization of fat for the production of respiratory substrate. In carbohydrate seeds the amount of fat present is usually small and is soon exhausted. The fall in the quotient in their case therefore only continues for a short period at the end of which the respiratory breakdown of carbohydrate again becomes the predominant process and consequently the quotient again rises towards unity. With fat-storing seeds a much longer and greater fall in the quotient occurs as fat continues to be converted into carbohydrate.

It will be observed from inspection of Fig. 1 that the

respiratory quotient of germinating flax seeds falls far below the theoretical value of about 0·7 for the oxidation of fat. This is due to the fact that as well as a complete oxidation of fat to carbon dioxide and water there is at the same time an accumulation of sugar in the germinating seeds. The source of this sugar must be fat, and oxygen is used in this transformation without any evolution of carbon dioxide taking place. Consequently the greater the ratio of sugar formation to complete oxidation the lower will be the ratio of carbon dioxide evolved to oxygen absorbed. It is very likely that in the oxidation of fat to carbon dioxide and water sugar is first formed, so that the respiratory quotients lower than 0·7 result from the breaking down of the sugar proceeding more slowly than its formation from fat.

In phase 3 an oxidation of carbohydrate substrate, produced as described under phase 2, takes place. This process goes on at the same time as the oxidation of the reserve fats, and its intensity is such that successively determined values of the respiratory quotient follow a rising course.

These theoretical views are further supported by the chemical analyses of germinating hemp seeds published by Detmer and given on page 21.

It will be seen that these three sets of figures roughly correspond with the three germination phases outlined above.

After seven days 15·56 gm. of fat have disappeared, and 8·64 gm. of starch have been formed, which corresponds closely with the respiratory behaviour during the second phase.

After ten days, it will be seen that the amount of fat has been still further reduced and at the same time some of the starch formed in the second phase has disappeared. Some of this starch has presumably been oxidized to water and carbon dioxide while some has probably been used in the formation of cellulose and protein, both of which substances have increased in quantity.

TABLE III

CHEMICAL CHANGES IN HEMP SEEDS DURING GERMINATION
(*From* Detmer)

	Fat	Starch	Protein	Undetermined compounds	Cellulose	Ash
100 gm. Ungerminated seeds	gm. 32·65	gm. —	gm. 25·06	gm. 21·28	gm. 16·51	gm. 4·5
After germinating 7 days	17·09	8·64	23·99	26·13	16·54	4·5
After germinating 10 days	15·20	4·59	24·50	26·95	18·29	4·5

When we come to consider the utilization of proteins by plants as respiratory substrates, we are faced with a serious lack of knowledge. Although the seeds of many species, particularly those of the Leguminosae, contain relatively large quantities of protein, there is no definite evidence that this protein is utilized in respiration. Analyses of some seeds of this type at progressive stages in germination have shown that the total amount of protein remains approximately constant, but in others, as for example those of the yellow lupin, *Lupinus luteus*, there is a very considerable loss of protein on germination, with a corresponding increase in simpler nitrogenous compounds such as amino-acids and asparagine, the amide of the amino-acid aspartic acid. Prianischnikov considered that this production of asparagine was connected with respiration. Asparagine has the formula $COOH.CH(NH_2).CH_2.CONH_2$ and it is calculated that the oxidation of protein to the final products asparagine, carbon dioxide and water would give a respiratory quotient of about 0·7. Bonnier and Mangin actually obtained a respiratory quotient of 0·58 for yellow lupin seedlings with radicles 4 to 5 cm. long and the cotyledons still closed, and a value of 0·42 with somewhat older seedlings with the cotyledons still closed but with radicles 8 to 9 cm. long. In still older seedlings with cotyledons half open and first foliage leaves visible, the quotient was recorded as 0·72. The low values of 0·58 and 0·42 strongly suggest the utilization of fat, part of which is converted to sugar and part ultimately to carbon dioxide and water, but there is nothing in these values of the respiratory quotient inconsistent with the utilization of a mixture of fat and protein or of a mixture of carbohydrate, fat and protein. The value of 0·72 found with older seedlings is near to that to be expected for the oxidation of protein to asparagine, carbon dioxide and water, but it could equally well be attributed to a utilization of fat. These values of Bonnier and Mangin referred to the late stages in germination. The authors determined the respiratory quotient of seeds of *Lupinus luteus* during early stages of germination and found values of the respiratory quotient for the first 24 hours of the germinating period between 1·0 and 0·9; only after 2 or 3 days did the quotient fall to about 0·76. Indeed, the changes in the respiratory quotient during germination of seeds of *Lupinus luteus* are very similar to those of a typical fat seed such as that of *Linum usitatissimum* (Fig. 1), and clearly the explanation of these changes in terms of material utilized can be the same for the two species. It will, however, be seen that no decisive conclusion can be drawn regarding the utilization of protein in the respiration of these seeds from values of the respiratory quotient alone.

It would appear that in some animal tissues the final oxidation products of protein are ammonia, carbon dioxide and water, and in plants, too, ammonia is to be regarded as the final nitrogenous end-product of protein decomposition. While the production of small amounts of ammonia during germination has been reported, there is no suggestion that this is at all common. If, however, protein were oxidized to carbon dioxide, water and ammonia the calculated respiratory quotient would be about 0·95. This is so near to that for carbohydrate that obviously respiratory quotients could provide no evidence of the utilization of protein in this way if carbohydrate were also present.

Evidence that protein may be used in the respiration of leaves starved by maintenance in the dark is more definite. Under such conditions the respiration rate gradually falls until it reaches a minimum constant level which is maintained for a considerable time if the leaves remain alive. This minimum constant respiration rate of starved leaves was termed by Blackman 'protoplasmic respiration' in contradistinction to 'floating respiration', in which carbohydrates or other reserve materials are utilized. Determinations of the respiratory quotient of starved barley leaves made by Yemm strongly suggest that the so-called protoplasmic respiration involves a utilization of protein. When the leaves are first placed in the dark the respiratory quotient is about unity, attributable presumably to the oxidation of carbohydrate. During the second day the quotient progressively falls to about 0·8; this could be due to the lessening utilization of carbohydrate and an increasing utilization of protein with the end-products asparagine, carbon dioxide and water. From the end of the second day up to the end of the fourth day the quotient remains in the neighbourhood of 0·8; Yemm supposes this is due to the continued lessening of the proportion of carbohydrate oxidized and an increase in the utilization of protein, of which an increasing percentage is converted to ammonia, carbon dioxide and water. In the final stage of starvation, after about 100 hours from the transference of the leaves to the dark, the quotient rises from about 0·8 to 0·9 as more and more of the protein is broken down to the end-products ammonia, carbon dioxide and water. Detailed chemical analyses of starved tobacco leaves made by Vickery, Pucher, Wakeman and Leavenworth also show that in the later stages of starvation amino-acids derived from the breaking down of proteins must be utilized in respiration, and in later work Yemm was able to isolate asparagine in crystalline form from starved barley leaves.

It has also been suggested that protein may be utilized as a respiratory substrate in *Asparagus* and spinach during storage. In *Asparagus* Platenius found the loss in sugar during three days in storage at 24° C. only accounted for about half the carbon dioxide given off. That the rest might have arisen from protein was suggested by the respiratory quotient which fell from 1·04 at the beginning of the storage period to 0·88 after two days in storage. This was followed by a rise to 0·95 after another day. From what has been written above it will be clear that these values are consistent with a utilization of a mixture of carbohydrate and protein. Values of the quotient down to 0·83 were found for spinach and Platenius suggested that here the leaf proteins might have been utilized in respiration. Similar values were obtained by Platenius for potatoes stored at 10° C. and 24° C. Bennett and Bartholomew, however, obtained the very low value of 0·49 for potato tubers stored at 7·5° C., while Platenius obtained values down to 0·45 at the beginning of storage of potato tubers at 0·5° C. Platenius regarded these low values as most easily explained on the basis of an incomplete oxidation of carbohydrate to organic acids.

Another possible instance of utilization of protein in respiration has been recorded by Stiles and Dent. When thin slices of red beetroot tissue were kept in aerated running tap water the respiratory quotient remained about unity for many days but with continued consumption of the substrate the value fell to 0·9 or less. This behaviour appears to be very similar to that observed in the other instances cited above, and a similar explanation is possible.

With regard to the cases we have just been discussing, the value of the respiratory quotient is chiefly considered from a relatively simple standpoint, namely, where substrates of highly complex chemical composition are broken down to simple compounds such as water and carbon dioxide. It may be assumed with a reasonable degree of certainty that such catabolic processes are useful to the plant mainly in so far as they set free energy. As has already been shown, the value of the respiratory quotient here depends upon the composition of the original substrate. There is, however, another aspect of the question to be considered. In the vast majority of plants, sugar may be taken as the fundamental substance from which by additive or subtractive processes the whole physical and physiological fabric of the organism is constructed. So far as we know, respiration is the universal accompaniment of life, and consequently, in the living plant, sugar is continually being utilized for the formation of other less complex or more complex substances. In the formation of these substances the sugar may be able to supply the exact amount of oxygen required, or on the other hand, additional oxygen may be needed, or surplus oxygen may be set free. In other words, the value of the measured respiratory quotient will depend upon the substance or substances that are being formed. A considerable amount of experimental evidence confirming this point is available, some of which will now be dealt with.

One of the first of such cases that suggests itself to us is that of the maturation of seeds which have fatty food reserves. We have seen that in the germination of such seeds, the breaking down of the fats first to sugars and later to carbon dioxide and water, results in a respiratory quotient of a value less than unity. As these fats, during maturation of the seed, are formed from sugars, one would expect, during maturation, the respiratory quotient to be greater than unity since the change from sugar to fat involves an elimination of oxygen. This has been experimentally demonstrated, as will be seen from the figures obtained by Gerber for linseed. During the maturation period immediately before ripening he found the average value of the respiratory quotient for six seeds to be 1·22, while during the germination of these same seeds the respiratory quotient fell to the average value of 0·64.

Then we have the often-cited case of succulent plants, in the cells of which accumulations of organic acids occur; malic acid in the Cactaceae and Crassulaceae, and oxalic acid in species of *Mesembryanthemum*. When these plants are placed in darkness, the amount of oxygen absorbed in the process of respiration is in excess of the amount of carbon dioxide evolved; in some extreme cases oxygen may be absorbed in marked quantities while no carbon dioxide is given off. Aubert, working with a species of *Opuntia* placed in darkness, obtained a mean value of 0·03 for the respiratory quotient. After the plants have been kept in darkness for a time, the accumulation of organic acids slows down and the rate of evolution of carbon dioxide gradually increases, with a corresponding increase in the value of the respiratory quotient, which, however, does not reach unity. A similar rise in the value of the quotient is produced by an increase in temperature. When the plants are exposed to sunlight, the acids are decomposed and carbon dioxide is set free. Thus there occurs an accumulation at night of organic acids, which, in the morning, are broken down with liberation of carbon dioxide. This latter gas is available for use in the process of assimilation. It has been suggested that this peculiar metabolism is beneficial to plants like succulents, in which, owing to the massive construction of their assimilatory organs, interchange of gases with the atmosphere may be relatively slow.

A somewhat similar case to that of succulents is that of plants whose leaves are coloured red by the presence of anthocyanin in their cells. Nicolas compared the respiration of green leaves with that of red leaves either from the same plants or from varieties of the same species whose leaves are red. He found that in every case examined the respiratory quotient of red leaves was lower than that of green leaves. He also found that these differences in the quotients were always due to a more active absorption of oxygen by the red leaves than by the green leaves. These differences in the quotients can apparently be related to a greater accumulation of organic acids in the leaves containing anthocyanin than in those from which this pigment is absent. The figures in Table IV, taken from Nicolas, show the amount of acetic acid

TABLE IV

RESPIRATORY QUOTIENTS AND ACID CONTENT OF GREEN AND RED LEAVES

(*From* Nicolas)

	Green leaves		Red leaves	
	Mg. Acetic acid	$\dfrac{CO_2}{O_2}$	Mg. Acetic acid	$\dfrac{CO_2}{O_2}$
Raphiolepsis ovata . . .	2·88	1·01	6·48	0·81
Photinia glabra	5·85	0·90	6·66	0·77
Acokanthera spectabilis .	8·21	0·94	11·11	0·71
Prumus cerasifera . . .	6·60	0·80	—	—
Prunus cerasifera var. Pissardi	—	—	10·80	0·70

(milligrammes per gramme fresh weight) in green and red leaves of four species examined, together with the respiratory quotients obtained.

Certain external physical conditions have been found to influence the value of the respiratory quotient. Temperature may markedly affect it in so far as it determines the velocity of the oxidation processes. In the already mentioned case of succulents, where increase in temperature results in the decomposition of organic acids, a marked increase in the value of the respiratory quotient is brought about. An increase in the value of the respiratory quotient with increase in temperature is recorded by Harrington for apple-seeds. An interesting fact which is probably connected with this, is that sugars and organic acids have been found by a number of workers to accumulate in dormant structures when they are stored at low temperatures.

If the concentration of oxygen in the atmosphere surrounding the respiring tissue is reduced below a given value, which varies with the plant material used, a marked rise in the value of the respiratory quotient results. This fact was clearly brought out by the researches of Stich in 1891. He found that the percentage of oxygen in the experimental atmosphere could be reduced from that of normal air, namely 20·8 per cent., down to values in the neighbourhood of 5 per cent. without bringing about any marked alteration in the respiratory quotient. When this lower limit of oxygen concentration, the exact value of which depended upon the species of the respiring plant, was passed, a marked increase in the quotient occurred. This is shown by the figures given in Table V, which are taken from Stich's paper.

Some values obtained more recently by Marsh and Goddard with thin slices of carrot root tissue show that in low oxygen concentrations the respiratory quotient progressively increases with reduction in the oxygen concentration (Table VI).

Increases in the concentration of carbon dioxide in the atmosphere surrounding the plant have a very marked depressing effect on the intensity of the respiratory processes, as will be seen later (p. 51). They also bring about a lowering of the respiratory quotient by causing a greater depression in carbon dioxide output than in oxygen intake. (See Table XI, p. 52.)

From the foregoing it thus appears that a study of the respiratory quotient may afford interesting clues as to the nature of the respiratory processes that are taking place within the plant. It has already been pointed out that, as many reactions may be safely assumed to be taking place in plant cells at one and the same time, it may be somewhat misleading to consider respiration as a simple physiological combustion involving the breaking

TABLE V

EFFECT OF OXYGEN CONCENTRATION ON RESPIRATORY QUOTIENT

(*From* Stich)

Experimental material	Percentage of oxygen in atmosphere	$\dfrac{CO_2}{O_2}$
Triticum vulgare, seedlings	20·8	0·98
	9·0	0·94
	5·0	0·93
	3·0	3·34
Zea Mais, seedlings	20·8	0·89
	9·0	0·96
	5·0	1·35
	3·6	1·37
Pisum sativum, seedlings	20·8	0·83
	9·3	0·86
	3·5	2·31
Narcissus poeticus, bulb	20·8	0·96
	10·2	1·04
	7·5	2·36

TABLE VI

EFFECT OF LOW OXYGEN CONCENTRATIONS ON THE RESPIRATORY QUOTIENT OF CARROT-ROOT TISSUES

(*From* Marsh and Goddard)

Percentage of oxygen in atmosphere	Oxygen absorbed in μl. per gm. per hr.	Carbon dioxide evolved in μl. per gm. per hr.	$\dfrac{CO_2}{O_2}$
5	53·0	43·5	0·82
5	56·8	43·8	0·77
2·5	45·0	56·3	1·25
2	40·5	52·0	1·28
2	37·8	45·8	1·21
2	28·3	49·0	1·73
2	28·3	55·6	1·96
1	15·6	52·0	3·3
1	14·2	49·5	3·5
1	19·2	66·5	3·3

down of a substrate to carbon dioxide and water with the absorption of oxygen. It is possible that more than one substrate may often, if not invariably, be involved; also it is possible that a number of reaction chains may simultaneously exist, giving rise to a diversity of final products. Each of these reaction chains will have its own particular value for the carbon dioxide-oxygen ratio, so that the respiratory quotient for a particular experimental subject, as measured by the experimental means at our disposal, will be the mean of all these reaction chain ratios. It may, as with values obtained from germinating fatty seeds, strongly indicate the nature of the predominant reaction chain that is taking place within the cells of the experimental tissue at the time of the experiment. On the other hand, where no one reaction chain is of sufficient intensity, or has a sufficiently characteristic carbon dioxide-oxygen ratio, to impress itself in an unmistakable way upon the observed respiratory quotient, the value of that quotient will convey little useful information as to what is happening inside the cells of the experimental material. Indeed, it may lead to entirely wrong conclusions; for example, the mean observed quotient resulting from a variety of reaction chains may have a result approaching unity and may consequently convey the idea that a simple complete combustion of carbohydrate substrate to carbon dioxide and water is the predominant chain. The value that is to be placed on such conclusions is obvious, even though it be supported on the part of the experimenter by the usual chemical arguments. These points are mentioned, not with a view to depreciating the existing results of various investigators, but by way of emphasizing the difficulties that confront workers in this field of research.

INTENSITY OF RESPIRATION

An examination of the various published general accounts of plant respiration reveals the fact that a considerable amount of vagueness exists where attempts are made to deal with the question of the intensity of respiration. In some instances modes of expression are used which are decidedly misleading; for example, one modern author of some repute refers to respiration intensity as respiratory energy, and states that the amount of carbon dioxide given off from a unit of living substance serves as a measure of this. Such a statement, besides being incorrect, involves a misuse of the term energy. Further, the complexity of the process of respiration and the intimacy of its relationship with all the vital processes of the living organism, do not always appear to be taken into account by writers when dealing with respiration intensity. This fact is evident when we consider the criteria most generally used for the purpose of expressing respiration intensity.

As pointed out in Chapter I, these criteria are naturally based on gaseous exchange. Usually either the amount of carbon dioxide evolved, or the amount of oxygen absorbed, by the respiring tissue is recorded. These quantities of gas are used as an indication of respiratory intensity by referring them to such quantities as the dry weight of the tissue used, the fresh weight of the tissue used, or the amount of nuclein nitrogen contained in the cells of the tissue.

When, however, we consider the already outlined variations that occur in the value of the respiratory quotient it is clear that estimation of respiration intensity based on the amount of oxygen absorbed, or on the amount of carbon dioxide evolved, may possess serious inaccuracies. It is true that such criteria, though far from being entirely satisfactory, are useful in an approximate and general way for comparative purposes. We are, however, faced with the fact that no really accurate method of expressing respiration intensity has, up to the present, been devised. This unfortunate state of affairs is due to the fact that in spite of the very considerable amount of research that has been carried out on respiration, our knowledge of the details of the process is still far from complete. We know that the process begins with some substrate and finishes with certain final products, and that in the process oxygen may be taken in and carbon dioxide given out. The actual respiration intensity, however, depends upon the rate at which the substrate is broken down by respiratory processes and upon the final products. In other words, the true measure of respiration intensity is the rate at which energy is set free, and this rate may or may not be accurately indicated by the rate at which oxygen is absorbed or carbon dioxide liberated. An experimental method, then, is required that will reveal the rate at which energy is set free as the respiratory substrate disappears, but we do not know with certainty what the substrate is. We know a little about it; for instance, we are reasonably certain that it is, in part at least, frequently a sugar or mixture of sugars. We know that other substances such as fats are frequently changed into sugars and the sugars broken down into simpler substances in the process of respiration. Here again, are we, strictly speaking, correct in considering the fat as the respiratory substrate, or should we consider the sugar as such? Moreover, there is evidence that proteins may act as substrates, though information on this point is very incomplete.

These facts should be kept in mind when studying the data at our disposal relating to respiration intensity and the factors which influence it. We will now proceed to a consideration of these data.

It is to be expected that as the various members of the vegetable kingdom differ so widely from each other morphologically, they also differ just as widely physiologically. Accordingly we find great differences between the respiration rates of different species. Also, as no two individuals of the same species are exactly similar morphologically, so no two such individuals exhibit exactly the same respiration intensity when subjected to similar external conditions. Amongst the most actively respiring plants are the fungi and bacteria. For example, Kostychev, working with a two-day-old culture of

Aspergillus niger on quinic acid at 16° C., found that it gave out 78·08 cubic centimetres of carbon dioxide per gramme of dry weight. Vignol found that a culture of *Bacillus mesentericus vulgatus* at 16° C. absorbed 48·51 cubic centimetres of oxygen per gramme of dry weight per hour. Some indication of the type of variation exhibited by the respiration intensity of different species of flowering plants is shown in the following table of values obtained by Aubert.

TABLE VII
RESPIRATION INTENSITY OF VARIOUS PLANTS
(*From* Aubert)

Plant	Temperature in ° C.	Vol. of oxygen absorbed per gramme fresh weight per hour
		cm.
Cereus macrogonus	12	3·00
Mamillaria cliphatidens . .	12	5·60
Sedum dendroideum . . .	12	19·00
Mesembryanthemum deltoides .	12	57·8
Lupinus albus	12	73·7
Vicia faba	12	96·6
Triticum sativum	13	291·00

In general, shade plants and succulent plants respire less actively than more normal types.

Then again we find that different parts of the same plant respire at different rates. In the higher plants, actively growing regions such as meristems and their adjacent immature tissues respire more actively than tissues which have reached their full development. Reproductive structures such as flowers show respiration intensities above the normal average intensity for the whole plant, and in the flower itself, the gynaecium and androecium respire more actively than the sepals and petals. A considerable amount of experimental data has been published by various workers proving these points, a few examples of which will now be considered.

Kidd, West, and Briggs measured the respiration of sunflower plants throughout their development from germination to maturity, and they also obtained comparative data of the respiration intensity of the different organs of mature plants. Their respiration intensities were expressed as milligrammes of carbon dioxide per gramme of dry weight of respiring cell-matter per hour

TABLE VIII

RESPIRATION INTENSITY OF VARIOUS PLANT ORGANS

(*From* Maige)

Species	Temp. in ° C.	Respiration intensity (c.c. CO_2 per gramme fresh weight per hour)				
		Sepals	Petals	Stamens	Pistil	Leaves
Verbascum thapsus . .	23·0	0·747	0·177	0·761	0·815	0·382
Penstemon gentianoides . .	23·5	0·571	0·398	0·602	0·689	0·300
Papaver rhoeas	22·0	0·390	0·367	1·041	0·690	0·332
Lavatera olbia .	22·0	0·615	0·303	0·576	0·894	0·394

at the temperature of 10° C., the external concentration of oxygen being that of the atmosphere. The respiration intensity so expressed they termed the *respiratory index*.

In Table IX (p. 37) are values taken from Kidd, West, and Briggs showing the respiratory indices of the various organs of sunflower plants of different ages.

Comparative data showing the relative respiration intensities of leaves and floral organs of four floweringplant species are given in the above table from results published by Maige in 1911.

VARIATIONS IN RESPIRATION INTENSITY DURING DEVELOPMENT

The variations in respiration intensity during the germination of seeds and during the early stages of the development of seedlings have received a considerable amount of attention from experimenters. De Saussure and a number of other pioneer workers recorded the fact that respiration, as measured by carbon dioxide output during the early stages of germination of seeds, showed a gradual increase with development. This work was carried further by Mayer in 1875 and Rischavi in 1876. These two workers observed the respiration of germinating wheat, the former measuring oxygen uptake and the latter carbon dioxide output. They both found that the respiration intensity increased from a very low value up to a maximum, and then gradually fell off in intensity.

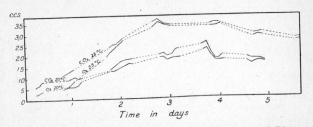

FIG. 2.—Curves showing changes in the respiration of *Pisum sativum* as measured by oxygen intake and carbon-dioxide output during the first five days of germination at temperatures of 20° C. and 25° C., respectively. In the upper curve respiration intensity is indicated as c.c. CO_2 or O_2 per 50 seeds per 3 hours, in the lower curve as c.c. CO_2 or O_2 per 50 seeds per 2 hours

(*After* Fernandes)

More recently, in 1923, Fernandes, using more precise experimental methods, examined the respiration rate of germinating peas, and the results of two of his experiments, conducted at 20° C. and 25° C. respectively, are shown in the curves given in Fig. 2.

It will be seen from this figure that Fernandes' results for *Pisum* agree with those of Mayer and Rischavi, the respiration intensity increasing fairly rapidly to a maximum value and then gradually decreasing. It has been suggested by Fernandes that the decrease after the maximum intensity has been reached is possibly due to exhaustion of available supplies of mineral salts.

The time during which this increase of respiration intensity up to a maximum value is taking place has been termed the grand period of respiration owing to the resemblance which exists between it and the grand period of growth. It is likely that respiration intensity largely runs parallel with growth-rate, but the two values are not always influenced in the same way by similar external conditions. For example, it has been shown in the case of germinating wheat that whereas the respiration intensity is higher at a temperature of 34° C. than at 23° C., the growth-rate is lower.

The authors found that the course of respiration of seeds of the sweet pea, *Lathyrus odoratus*, during germination in the dark depended on whether the testas were present or not. If they were present five phases could be distinguished, namely, first a fairly rapid increase in the rate of respiration as the seeds absorbed water, next a period during which the respiration remained constant which continued for a variable time until the seed coats were ruptured, thirdly a rapid rise in the rate following the rupture of the coats followed by a fourth phase during which the rate remained at a maximum, and finally a phase during which the rate slowly fell. If the seed coats were first removed the second phase was eliminated so that the respiration rate rose continuously to a maximum after which there was the final fall in the rate. The authors concluded that this final fall in rate is related to the conditions of experimentation which tend towards a reduction in the rate of transpiration and so the conveyance of respirable material from the cotyledons to the growing points where respiration is most active. In the opinion of the authors it is therefore at least possible that the conception of a grand period of respiration during the germination of seeds may result from the conditions of experimenta-

TABLE IX

CHANGES IN RESPIRATORY INDEX OF SUNFLOWER PLANTS WITH AGE
(From Kidd, West, and Briggs)

Days from germination	No. of plants used	Dry weight of a single plant	Respiratory index (mg. CO_2 per gm. dry weight per hour) of—					
			Entire plant	Stem	Leaves	Stem apex	Total inflorescences	Flowers on lateral shoots
1	30	0.0225	2.90	—	—	—	—	—
2	25	0.0223	3.00	—	—	—	—	—
4	25	0.0242	3.00	—	—	—	—	—
13	10	0.1009	2.80	—	—	—	—	—
22	8	0.630	3.00	(3.00)	(3.00)	—	—	—
29	2	4.065	2.30	—	—	3.00	—	—
36	1	12.85	1.21	0.81	1.56	—	—	—
43	1	22.05	1.03	0.69	1.38	—	—	—
50	1	45.15	0.94	0.46	1.52	2.56	—	—
59	1	93.20	0.66	0.33	1.32	1.78	—	—
64	1	98.30	0.71	0.34	1.24	—	—	—
89	1	294.7	0.48	0.31	0.90	1.13*	1.13	1.13
99	1	377.4	0.37	0.25	0.45	0.89	1.04	0.95
112	1	818.3	0.26	0.098	0.375	0.75	0.85	0.97
136	1	419.5	0.39	0.081	0.44	0.96	0.965	

* From this date onwards the stem apex was the inflorescence only.

tion and may not occur with seeds germinating in the open.

When considering the respiration of more mature plant organs, we find that the intensity of the process tends to decrease with the age of the organ in question. This point has been clearly brought out by Kidd, West, and Briggs in their work with the sunflower. Table IX summarizes their observations in this connexion and shows how the respiratory index (see p. 34) decreases with age in the various individual parts of the plant.

Other investigators working with flowers, stems and leaves of various plants, generally speaking, record a similar decrease in respiration intensity with age.

As a result of more recent work dealing with the relation between respiration rate and the age of plant organs, a further point has revealed itself. In some leaves and fruits examined, the already mentioned decrease in respiration intensity with age continues apparently until the structure is fully developed. The organ then passes into a senescent phase of its existence, during which the respiration intensity increases up to a maximum value, after which it again decreases.

Dealing with this are researches carried out by Blackman and Parija in 1928 and by Kidd and West in 1930 on the respiration of stored apples. The course of the respiration intensity of apples during storage is shown in Fig. 3, taken from the paper by Kidd and West. In the three curves shown in this figure which indicate the respiratory activity of stored Bramley's Seedling apples at different temperatures, the respiration intensity, in each case, is seen to increase to a maximum value and then to decrease. The increase of respiratory activity to a maximum was called by Kidd and West the climacteric rise. These results have been confirmed for apples by subsequent workers while similar results have been obtained by Gustafson for the tomato, Olney and Wardlaw and Leonard for the banana, Roux for peaches and plums, Pratt and Biale for the avocado, and by Kidd,

West, Griffiths and Potter for pears. According to Biale and Young climacteric rise is not shown by citrus fruits respiring in normal air, but does occur in atmospheres containing 34, 68 or 99 per cent. oxygen.

A tentative theory was put forward by Blackman and Parija as a possible explanation of this behaviour of respiration of stored fruits. The essence of this theory is that during the senescent phase the protoplasmic control

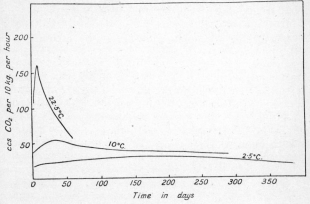

FIG. 3.—Curves showing the course of respiration intensity of Bramley's Seedling apples during storage at temperatures of 22·5° C., 10° C., and 2·5° C.

(*After* Kidd and West)

of hydrolysis, which they term 'organization resistance', is weakened, with the result that hydrolysis increases and produces an increased amount of substrate for respiration. This increased amount of substrate results in a rise in the respiration intensity. Later, the starvation factor comes into operation and the supply of substrate available for hydrolysis diminishes, with the result that the respiration intensity decreases, thus causing the later downward slope of the curve from the maximum shown in Fig. 3.

Subsequently, after examining the chemical changes taking place in stored apples, Kidd gave this theory greater precision. He suggested that the climacteric rise resulted from an increase in the amount of 'active' fructose, the supposed respiratory substrate, in the protoplasm. This was supposed to result from a change in the permeability of the inner plasmatic membrane so that normal inactive fructose, which had accumulated in the vacuole, passed into the protoplasm where it was converted into active fructose. In pears there occurs a further increase in respiratory activity after the climacteric rise. This accompanies breakdown of the flesh of the fruit and is followed by a final fall in respiration rate which in absence of attack by micro-organisms would ultimately reach zero. This last rise would be explicable, in Blackman's terminology, as due to a final breakdown in cell organization and so of organization resistance.

Two other factors that may be termed internal, and which influence the intensity of the respiration process in plants, are water content and seasonal periodicity. Various workers have investigated the effect on respiration of changing the water content of the cells of vegetative organs of plants. In some instances this change in water content was brought about by simple desiccation, in others by the osmotic action caused by immersion of the tissue in distilled water or solutions of glucose of different concentrations. In the majority of plants experimented upon in this way, it has been found that, up to a point, increase or decrease in water content brings about an increase in respiration intensity, increase in water content causing greater increase in respiration than decrease in water content. Amongst the exceptions to this rule are *Asparagus* and paeony which show no increase in respiration rate when their water content is reduced. In every case, as one would expect, desiccation, if carried beyond a certain limit which varies with different species, brings about a reduction in respiration intensity which continues to zero as complete desiccation and death of the tissues is reached. In potato tubers it has been found that withdrawal of water results in a decrease in respiration rate; a probable parallel to this is the decrease in respiration intensity exhibited by seeds during the process of ripening.

A somewhat similar case in which plant tissues lose water and pass into a resting stage is that of mosses and lichens. Many of these plants are able to withstand extreme desiccation while still remaining alive, and are able, when opportunity favours, to re-absorb water and resume their vital activities. A number of workers have investigated the relationship between water content and respiration rate in various moss species, and have shown that the two values always vary in the same direction. By way of example, the following set of figures obtained by Jönsson for *Mnium undulatum* may be taken.

TABLE X

EFFECT OF WATER CONTENT ON RESPIRATION INTENSITY OF *MNIUM UNDULATUM*

(*From* Jönsson)

Percentage water content	Carbon dioxide evolved per gramme dry weight per 10 hours c.c.
40	0·750
59	1·350
65	3·900
84	9·680

The increase in respiration intensity that is observed in seeds during the period in which they are absorbing water preparatory to germination is a further instance similar to those discussed above.

The effect of seasonal periodicity on the intensity of respiration presents certain features of interest. Bonnier and Mangin in 1885 found that in perennial plants growing in temperate climates the average respiration was greatest in spring, and showed a slight decrease in summer; a more rapid decrease to a minimum value occurred with the onset of winter, after which the intensity again increased with the return of spring. In the course of this yearly cycle, two subsidiary maxima were observed, the first being related to the expansion of new leaves in the spring, and the second appearing at the time of flower production.

Attention has more recently been drawn to the question of seasonal effect by the experiments of Inamdar and Singh carried out at Benares on the respiration of the leaves of *Artocarpus integrifolia*. It will be noted here that the plants are growing under tropical climatic conditions. In this region spring occurs at the end of March, during which period the plants produce new leaves and shoots. Summer quickly follows spring and is very hot and dry until the end of June, when monsoon rains begin and continue until the end of September. After the rainy season comes a comparatively dry autumn of about two months' duration, followed by winter, with a relatively low temperature and only very occasional rain.

The general course of respiration of the leaves of *Artocarpus* during the year was found to be as follows: at the beginning of summer, respiration intensity falls to a minimum value which persists until the coming of autumn, it then gradually increases to a maximum level which continues through the winter and spring, again falling as spring gives place to summer.

A striking difference was found to exist between the respiratory behaviour of leaves collected during the summer and those collected during the winter. When leaves were kept in darkness, and their respiration rates measured over periods of several days, leaves collected in winter gave the typical starvation curve in which respiration intensity first rapidly decreased as carbohydrate reserves were depleted until a steady low respiration level was reached and maintained, that is, the change from the 'floating' to the 'protoplasmic' types of respiration of Blackman was observed. In leaves collected in summer and similarly treated, the initial phase was found to be absent, and instead a low, almost uniform, respiration rate was maintained throughout the experiment.

It would appear that the difference between the respiratory intensities of the two seasons might be related in some measure to the relative abundance of respirable carbohydrate substrate, as photosynthetic activity is also at a minimum during summer. More complex causes, however, underlie the matter, as is pointed out by the authors of the work. Some factors apparently bring about a depression of the activity of the metabolic mechanism of the cells of the plant during the summer, this depression affecting both respiration and assimilation alike.

Water content cannot be the determining factor as this is practically the same in summer as in winter. Growth activity also bears no relationship to respiratory activity as the former is practically at a minimum while the latter is at a maximum, that is, during autumn and winter; active growth takes place in early spring and during the rains of late summer. As regards temperature, we find a contrast between plants of temperate regions and those of Benares, in that in the former, the minimum respiration intensity is related to the lowest temperature.

THE EFFECT OF EXTERNAL FACTORS ON RESPIRATION INTENSITY

The various factors that we have so far considered, that influence the intensity of the respiratory processes of plants, are more or less entirely dependent upon conditions within, or upon the specific nature of the living protoplasmic system of, the cells. In addition to these are a number of environmental factors that are found to influence plant cells directly or indirectly in such a way as to bring about changes in their respiration rates. These factors may be generally termed external factors, the chief of them being temperature, light, changes in the composition and pressure of the external atmosphere, and the introduction of various chemical compounds into the respiring cells. The investigation of a number of these factors has received a considerable amount of attention, such work being materially helped by the fact that they can be accurately controlled.

Temperature.—Attempts have been repeatedly made to analyse the effects of temperature changes on the respiratory processes. Some investigators have gone to the extent of formulating mathematical laws, which are based on experimental data of varying reliability, connecting temperature with respiration intensity. In the present state of our knowledge such laws cannot be treated with any confidence as regards their validity, and their discussion here would be of little help. In fact, the only generalization that we can so far make with any measure of certainty regarding the point in question is that, within certain limits, increase in temperature results in increase in respiration rate.

The investigations dealing with the effect of temperature changes on respiration intensity, generally speaking, fall into two categories, namely, those dealing with rapidly developing structures and actively functioning organs, such as seedlings and leaves, and those dealing with resting or senescent organs such as tubers and fruits. Seedlings, owing to the fact that for many reasons they form admirable subjects for experiment, have received considerable attention from time to time. A good deal of the earlier work was carried out with a view to ascertaining the optimum temperature for respiration, but so far, no satisfactory conclusions have been reached with regard to this. Although increases in temperature up to values in the neighbourhood of 45° C. are accompanied by corresponding initial increases in respiration intensity, these high initial rates are not always maintained. This point is well illustrated in Fig. 4, which is taken from the work of Fernandes and shows the effect on the respiration intensity of four-day-old pea seedlings, of changing the temperature from an initial value of 25° C. to various other experimental values. The operation of the time factor is well shown in this figure; it will be seen that raising the temperature of the seedlings to values higher than about 30° C. to 35° C. results in a falling off in respiration rate with time from the initial maximum rate for the temperature under consideration. For temperatures between 0° C. and 45° C., increase in temperature results in an increase in this initial respiration intensity,

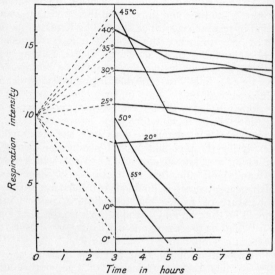

FIG. 4.—Curves showing the comparative rates of respiration of four-day-old seedlings of *Pisum sativum* as affected by different temperatures and related to time

(*After* Fernandes)

but temperatures above 45° C. result in a progressive lowering of the initial rate (cf. curves for 50° C. and 55° C. in Fig. 4). It would appear then that we must probably consider a temperature in the neighbourhood of 30° C. as the optimum for the experimental material under consideration, as at this temperature there occurs no falling off in the respiration intensity with time. A complication entering into work of this kind, when seedlings in the early stages of their development are used, results from the fact already described on page 35, that the respiration intensity of seedlings germinating under certain conditions does not follow a level course, but shows an initial rise and subsequent fall. Owing to the changes in germination rate produced by different temperatures, the time and duration of this so-called 'grand period of respiration' will be affected by temperature, so that it may influence the form of temperature-effect graphs similar to those shown in Fig. 4, causing upgrades or down-grades which are apt to be erroneously interpreted. This has probably happened with some published work on the lines under discussion.

Another aspect of the question is that involving determinations of the temperature coefficient of the respiratory process. Such determinations are made with a view to exploring the possibilities of the connexion between respiration and known chemical reactions which obey the Van't Hoff rule, that is, reactions, the velocity of which is approximately doubled or trebled by a rise in temperature of 10° C. The temperature coefficient is denoted by the symbol Q_{10} and is the ratio of the rate of the reaction or process at one particular temperature to its rate at a temperature 10° C. lower.

The operation of the time factor as outlined above often renders the determination of temperature coefficients for respiration, with any degree of certainty, very difficult; in fact, at temperatures above about 30° C. attempts to make such determinations are probably of doubtful value. At temperatures between 0° C. and 30° C., in material where the respiration intensity remains reasonably constant, estimations of the temperature coefficient may be of considerable value. In this connexion we have a Q_{10} value of 2·5 obtained by Clausen for wheat, lupin seedlings and *Syringa* flowers between 0° C. and 20° C., and 2·1 by Blackman and Matthaei for cherry-laurel leaves over a range of 16° C. to 45° C.

Gerhart working on the respiration of strawberry fruits obtained a Q_{10} value of 2·5 between 5° C. and 25° C., but with temperatures above 25° C. it was impossible to arrive at any consistent value for the coefficient. Values of the same order of magnitude for the temperature coefficient of the respiration of germinating peas between 0° C. and 20° C. are indicated by the data of Kuijper and Fernandes.

An interesting effect of temperature upon respiration intensity is that described by Müller-Thurgau and

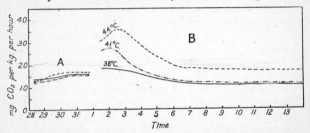

FIG. 5.—Curves showing the course of respiration intensity of potato tubers at a temperature of 19° C., at A before and B after they had been kept for eight hours at 38° C., 41° C., and 44° C. respectively. The experiment extended from 28 March to 14 April

(*After* Müller-Thurgau and Schneider-Orelli)

Schneider-Orelli for the potato. They investigated the changes in the course of carbon dioxide output by potato tubers brought about by maintaining them at various temperatures for eight-hour periods, and then recording their rates of carbon dioxide evolution over periods of fourteen days at a temperature of 19° C. The effect of this treatment is indicated in Fig. 5. It will be seen that the previous heating produces an initial increase in respiration intensity which subsequently falls off. In those tubers which were heated to temperatures higher than 38° C. the falling off ceases while respiration intensity is still higher than it was before the heating was carried out. In other words, heating potato tubers for eight hours at temperatures of 41° C. and 44° C. produced a permanent increase in respiration intensity at normal temperatures.

The effect of temperature on the respiration of stored apples formed part of the already mentioned investigation of Kidd and West (cf. p. 38). The three curves given in Fig. 3 show the course of the respiration intensity during the period of senescence of apples when stored at temperatures of 2·5° C., 10° C., and 22·5° C. respectively. These investigators found that one effect of increased temperature was a shortening of the senescent period beginning with the removal of the fruit from the tree and ending with their death from fungal attack. Another effect was an increase in respiration intensity (see Fig. 3). This influence of temperature on senescent drift and carbon dioxide output was found to be such that the total amount of carbon dioxide liberated from the time of gathering to the time of death was approximately the same, no matter at what temperature the fruit was stored. It was also found that the amount of dry matter lost by the fruit during the senescent period was approximately constant, regardless of temperature.

Light.—Various attempts have been made from time to time to determine the effect of light on respiration intensity, the results of which are inconclusive or contradictory. The present position, therefore, with regard to this question is that light appears to have either very little or no direct effect on respiration. Indirectly, however, in plant organs containing chlorophyll, light may have a very considerable effect on respiration intensity through its influence upon the supply of respiratory substrate resulting from photosynthesis. Light may have an apparent effect on respiration rate owing to its action in causing a decomposition of organic acids with a resulting liberation of carbon dioxide, as mentioned in connexion with the metabolism of succulent plants on page 26.

Concentration of Oxygen.—Although the normal

respiration in plant organs is dependent upon an adequate supply of oxygen for its continuance, until recently the widely accepted view was that considerable variations in the concentration of this gas in the surrounding atmosphere might occur without causing any change in respiration intensity. For low concentrations, figures were put forward by Stich indicating that the percentage of oxygen in the atmosphere might be reduced to a low value before a change to anaerobic respiration was indicated by the already mentioned rise in the respiratory quotient (see p. 28). As regards the effect of oxygen concentration on the intensity of respiration, different tissues appear to exhibit different behaviour. Denny working with wheat seedlings found that reduction in the oxygen concentration from that of normal air to the range 12·6 to 14·4 per cent. had no effect on carbon dioxide production but caused a reduction of oxygen consumption of about 5 per cent. Further reduction, however, in oxygen concentration to the range 9·6 to 10·4 per cent. resulted in the carbon dioxide output being reduced by about 2·9 per cent. Among whole storage organs, Denny found that with potato tubers, reducing the oxygen concentration from 20·9 per cent. to 13·3 per cent. over a period of 25 hours had no effect on carbon dioxide output, while, on the other hand, oxygen intake was reduced by 4·2 per cent. Reduction of the oxygen concentration to 2·4 per cent. over a period of 40·5 hours depressed carbon dioxide output by only 1·73 per cent., while oxygen intake was depressed by 13·8 per cent. With Jerusalem artichoke tubers depletion of oxygen concentration to 14·8 per cent. in 24 hours had no effect on carbon dioxide production but depressed oxygen intake by 5·4 per cent. Depletion of oxygen concentration to 11·8 per cent. in 16·5 hours resulted in a reduction of carbon dioxide output by 1·55 per cent. and of oxygen intake by 5·6 per cent. Choudhury found that variation in oxygen concentration between 6·2 and 100 per cent. had no appreciable effect on carbon dioxide output by potato tubers, but the rate of carbon dioxide evolution in pure nitrogen was much less. With artichoke tubers, on the other hand, there was observed a progressive lowering of carbon dioxide output with reduction in oxygen concentration to 10·7, 6·7 and 3·5 per cent., but in 100 per cent. oxygen the rate of respiration was the same as in air. Variable results were found with carrot roots; with one root carbon dioxide output in 3·5 per cent. oxygen was lower than that occurring in air, while in another root it was higher, and in pure nitrogen the rate of carbon dioxide might exceed that in air. In atmospheres containing a higher concentration of oxygen than that of this gas in air the respiration rate increased with increase in oxygen concentration so that the maximum rate occurred in 100 per cent. oxygen. It would thus seem that with whole carrot roots there is a minimum value of oxygen concentration above and below which carbon dioxide output increases. Marsh and Goddard came to a similar conclusion regarding the evolution of carbon dioxide from thin slices of carrot root in low oxygen concentrations, the rate of carbon dioxide output increasing with lowering of the oxygen concentration below 5 per cent. The rate of oxygen absorption falls, however, with lowering of the oxygen tension, so that the respiratory quotient increases from 0·82 in 5 per cent. oxygen to 3·5 in 1 per cent. oxygen. Thus in low oxygen concentration only a part of the respiration is aerobic, and whether carbon dioxide output rises or falls with diminishing oxygen tension must depend on the two processes of aerobic respiration and anaerobic respiration and the way in which each is affected by oxygen. This matter will be dealt with in subsequent chapters. Carrot root tissue thus appears to behave like that of the fruit of the apple in which Blackman found that in pure nitrogen the carbon dioxide production was high, and that as oxygen was admitted it fell off rapidly until the oxygen concentration was about 5 per cent. to 9 per cent. Further gradual increase in the oxygen concentration up to 100 per cent. resulted in a corresponding steady increase in carbon dioxide output.

It is interesting to note that seedlings of wheat and of rice behave differently in low oxygen concentrations. According to Taylor, although in both species the rate of oxygen absorption falls progressively as the oxygen concentration is reduced from that of air to zero, in wheat the rate of carbon dioxide output decreases while in rice it increases with lowering of the oxygen concentration, so that in pure nitrogen the rate of carbon dioxide output is about 50 per cent. higher than in normal air. It is suggested that the ability of rice to grow better than wheat under conditions of poor aeration is related to its capacity to respire under anaerobic conditions.

Concentration of Carbon Dioxide.—Increased concentration of carbon dioxide in the atmosphere brings about very marked depression in the respiratory process. Although it has been known since the time of de Saussure that high carbon dioxide concentrations in the surrounding air are injurious to plants, the effect of this factor on plant respiration was not clearly brought out until the publication of the researches of Kidd. He examined the effect of various concentrations of carbon dioxide in air, on both the oxygen intake and carbon dioxide output of germinating seeds of white mustard, and also on the carbon dioxide output of cherry-laurel leaves. For germinating mustard seeds, the data resulting from Kidd's experiments are set out in Table XI.

It will be seen that the depressing effect of concentrations up to 50 per cent. carbon dioxide varies roughly with the square root of the concentrations.

With regard to the experiments upon leaves in this connexion, owing to technical difficulties the results obtained are less conclusive, but from the data obtained it appears that the inhibiting effect of high carbon dioxide concentrations is confined to the 'floating' respiration, the 'protoplasmic' respiration being unaffected. Livingston and Franck, however, reported that the effect of high carbon dioxide concentration on the respiration rate of leaves of *Hydrangea otaksa* depended on the time of year, being least in December and highest in April. In the former month the respiration in the dark was about the same in air, in 5 per cent. carbon dioxide and in

TABLE XI

THE RETARDING INFLUENCE OF INCREASED CONCENTRATIONS OF CARBON DIOXIDE UPON THE RATE OF NORMAL RESPIRATION IN GERMINATING WHITE MUSTARD SEEDS, MEASURED BY CO_2 PRODUCTION AND OXYGEN CONSUMPTION

(*From* Kidd)

	Concentrations of carbon dioxide initially present					
	0%	10%	20%	30%	40%	80%
After 14 hours:						
c.c. CO_2 gain	58	48	38	33	26	17
c.c. O_2 loss	71	57	49	45	38	32
Respiratory quotient	0·82	0·84	0·77	0·73	0·69	0·53
After 40 hours:						
c.c. CO_2 gain	173	158	96	75	61	41
c.c. O_2 loss	197	185	122	104	97	90
Respiratory quotient	0·87	0·85	0·75	0·72	0·63	0·45

Conducted in dim diffuse light. 20 per cent. oxygen present initially in each experiment. 15 gm. of seed set dry on 50 c.c. damp sand and 10 c.c. tap water in each experiment. Results obtained from analyses. Temperature of experiments, 25·5° C. by thermostat.

20 per cent. carbon dioxide, whereas in April the respiration rate in 20 per cent. carbon dioxide was very low. A period of illumination generally had the effect of lowering the respiration rate in this high concentration of carbon dioxide.

Kidd also experimented with green peas, and suggested that the dormancy of certain seeds is brought about by the presence of high carbon dioxide concentration in the tissues resulting from restriction by the testa of the free passage of gases. It would appear, however, that the whole question of the cause of dormancy requires further investigation in view of the researches of Thornton and Denny. Thornton found that freshly harvested potatoes sprouted in seven days when kept under moist conditions in an atmosphere containing 5 or 10 per cent. of oxygen. On the other hand, they remained dormant for 47 days when the atmospheric oxygen concentration was maintained at 20 per cent. Denny found that the internal atmospheres of *Gladiolus* corms that had been kept in a dormant condition in moist soil at room temperature for periods of 6 to 18 months, contained only 3·8 per cent. carbon dioxide and 18·9 per cent. oxygen. During this storage period the respiration rate as measured by carbon dioxide output was low. On removing the corms from the soil the respiration rate rose, reaching a maximum in from 20 to 30 hours, after which it fell back to its initial low level. At the period of maximum respiration the internal atmospheres of the corms contained 30 per cent. carbon dioxide and 7 per cent. oxygen. It thus appears that the low respiration rate characteristic of dormant tissues is not necessarily accompanied by a high concentration of carbon dioxide in the intercellular spaces of such tissues.

Ionized Air.—Before leaving the question of the changes in respiration intensity produced by variations in the composition of the atmosphere, one other point seems worthy of mention, namely, the effect of ionized air. It has been found that plants, or parts of plants, respire more actively in air that has been ionized by means of radio-active substances than they do in air that has not been so treated. As the gases of the atmosphere are to some extent ionized during daytime by the action of sunlight, this factor probably has some small effect on the respiration of plants growing under natural conditions.

Sugar.— The concentration of sugar in culture

solutions in which fungi are growing has a marked effect on the respiration intensity of these plants. This fact has been demonstrated by Maige and Nicolas and by Kosinski. Various sugars were used in the experiments of Maige and Nicolas, who found that, generally, respiration intensity increased with increased sugar concentration up to a point when plasmolysis set in, when the respiratory rate decreased. It has also been shown that the respiration of etiolated leaves which are poor in sugar, is considerably increased by immersing the petioles of the cut leaves in sugar solution. Also Palladin found that the respiratory activity of starved leaves of *Vicia Faba* was greatly increased by floating them on a solution of sucrose. Hanes and Barker also concluded that the respiratory activity of potato tubers was directly related to the concentrations of sugars.

Inorganic Salts.—Much work in recent years has indicated that the intensity of respiration is affected by inorganic salts in the external medium. Lundegårdh and Burström found that the respiration of the roots of wheat seedlings was increased in the presence of dilute solutions of a number of chlorides. The resulting increase in respiration rate they called 'anion respiration' as they found that its value was related to the total amount of anion absorbed by the roots. A number of other workers have recorded an increase in respiration rate related to the presence of inorganic salts. Thus Steward and Preston found that the chloride, bromide, nitrate and sulphate of potassium all brought about an increase in the respiratory activity of thin slices of potato tuber tissue, whereas the chloride, bromide and nitrate of calcium all brought about a decrease in the intensity of respiration. Bennet-Clark and Bexon, on the other hand, found that calcium chloride as well as potassium chloride, brought about an increase in the respiration rate of thin slices of red beetroot. Robertson found the chlorides of potassium, sodium, lithium, calcium and magnesium all increased the respiratory activity of thin slices of carrot root, but that while this effect was maintained with salts with monovalent cations for many hours, with calcium and magnesium chlorides the level of the respiratory activity soon fell to that of tissue in distilled water or even lower.

Lundegårdh and Burström consider that in presence of salt solutions the total respiration is made up of two independent components, a fundamental respiration unconnected with salt and an anion respiration closely connected with the absorption of salt into the cell against a concentration gradient. They consider that the fundamental respiration involves a catalysis system with manganese while an iron catalysis system is concerned with the anion respiration system. In support of this view is the fact that cyanide inhibits both the anion respiration component of respiration and the accumulation of salt, but does not affect the fundamental respiration.

The term salt respiration was proposed by Robertson for the component of respiration related to the absorption of salt. The reason why this increase of respiration resulting from the presence of salt is related to the absorption of anion is due to the fact that when tissue is placed in a salt solution two processes generally occur: an interchange of cations between tissue and external medium and an accumulation of salt, that is of both ions, in the tissue. If the salt respiration is related to the latter process and not to the former it will thus be proportional to the amount of anion, but not to the amount of cation, that passes from the medium into the tissue.

Acids.—It is possible that dilute acids may have an effect similar to that of salts. Wehner, measuring intake of oxygen, observed that the aquatic moss *Fontinalis* respired more actively in a mixture of 0·0001 N nitric acid and 0·1 N sodium nitrate than in corresponding solutions of either of these individual compounds.

Various Organic Substances.—The effects of various organic poisons on the respiratory processes of plants are of interest. The effect of chloroform upon the respiration of cherry-laurel leaves was investigated by Miss Irving, who found that small doses caused an increase in respiration intensity which might persist so long as the application of the substance was continued. Medium doses caused an initial increase in intensity which was followed by a decrease to much below normal, the larger the dose the more rapid the decrease. Strong doses of chloroform resulted in a rapid fall in respiration rate to zero without any initial increase occurring. Other workers have carried out similar investigations with various plant organs and using a variety of poisons including ether, cocaine, morphine, quinine, chloral hydrate, caffeine, ethyl bromide, formaldehyde, acetone and ethyl alcohol. Broadly speaking, the results described are very similar to those outlined above. Although vague theories have been advanced in attempts to account for these effects of poisons on plant cells, no generally satisfactory conclusions have been so far reached.

Since undoubtedly a number of enzymes are involved in respiration it is to be expected that any substance which inhibits an enzyme action forming part of the respiratory process will also inhibit respiration. As will be shown in a later chapter, the effect of various enzyme inhibitors on respiration has been used as an argument in theories on the respiratory mechanism.

The Effect of Mechanical Stimulation on Respiration.—While engaged in an examination of the respiration of leaves of the cherry laurel, Audus found that if the leaves were removed from the chamber containing them and then replaced after a few minutes, their respiration was considerably higher than before removal. Stroking or bending the leaf had the same effect. A number of other leaves examined subsequently behaved similarly, the increase in respiration rate varying from about 20 to 183 per cent. In an atmosphere of nitrogen the treatment had no effect on the rate of carbon dioxide evolution so that the effect would appear to be related to an oxidation phase in the whole respiration process. An increase in the rate of respiration as a result of handling was also observed by Barker in potato tubers.

If leaves were subjected to a series of such treatments the increase in respiration rate was progressively less with each successive one. The handling, stroking or bending thus appears to act as a mechanical stimulus and can reasonably be described as mechanical stimulation, but like the action of so many stimuli the mechanism of the action is unknown.

It may also be mentioned that other forms of stimulation have been found to cause an increase in respiration intensity. As examples of this we have the increase in respiration rate exhibited by the carpels of flowers after pollination has taken place, and similar increases in roots undergoing curvature as a result of geotropic response.

The Effect of Wounding on Respiration.—It has long been known that when a plant organ is wounded an increase in respiration intensity results. Böhm in 1887 called attention to the fact that when potatoes are cut they exhibit an increased output of carbon dioxide. In 1891, Stich published a more comprehensive account of the phenomenon, having investigated it in other plants in addition to the potato. He also showed that when a potato was cut into two parts, and the cut surfaces joined together again by means of neutral gelatin, the resulting increase in respiration intensity was less than when the cut surfaces were left exposed to the air. Five years later Richards described wounding experiments upon potatoes, carrots, beet, and the hypocotyls and roots of *Vicia* and *Cucurbita*, and various leaves. In all these he obtained, after injury, respiration rates which varied in intensity and duration with the character of the tissue involved and with the extent of the wounding. This increased respiratory activity usually reached a maximum within two days, after which it fell gradually until an approximately normal rate was resumed. In bulky tissues, for example potato and carrot, there occurred during the first two or three hours a sudden increase followed by a rapid decrease in the amount of carbon dioxide evolved. This was due to the escape from the cut surfaces of gas previously enclosed in the tissue.

In order to investigate the cause of this increase in respiration intensity which follows wounding, Hopkins measured the respiration of cut potatoes and also determined the variations in sugar content which occurred in the cut tuber. He found that this increased by from 53 to 68 per cent. of the original sugar content, and that the maximum occurred several days after wounding, after which it slowly fell. He also found that the maximum sugar content was reached after the time of maximum respiration intensity had been passed; this appeared to be due to the fact that suberization of the wound occurred, causing accumulation of carbon dioxide in the tissue, thereby bringing about a lowering of the respiration rate. It has already been suggested that accumulated carbon dioxide in tissues may frequently be a limiting factor in respiration intensity.

CHAPTER III

ANAEROBIC RESPIRATION

IT is well known that most animals, when deprived of oxygen, cease to respire and die in consequence. The behaviour of plants is in marked contrast to this. When removed from a supply of oxygen a plant which normally respires aerobically continues to give out carbon dioxide, and the production of this gas may continue for a longer or shorter time, according to the plant material. If anaerobic conditions are maintained for too long a time the plant suffers injury and may be killed in consequence, but if the absence of oxygen is not too prolonged, the plant, on return to a normal atmosphere, behaves quite normally and is found to be perfectly healthy. The actual time for which an aerobic plant can withstand anaerobic conditions without injury depends upon various conditions such as temperature and food supply. It was reported by Chudiakow in 1894 that maize seedlings in absence of oxygen die in 24 hours at 18° C. and in 12 hours at 40° C., while it has been stated that apples and pears remain uninjured for months in an atmosphere of pure nitrogen or pure hydrogen.

The first definite observation on record of the evolution of carbon dioxide by plants in absence of oxygen was made in 1797 by William Cruickshank, Chemist to the Ordnance and Surgeon of Artillery. We have thought it of interest to quote his own description of one of his experiments, for it was none other than the ordinary laboratory method of demonstrating anaerobic respiration which has been performed and observed by many thousands of students in succeeding decades, although pea seeds are usually substituted for barley grains.

'*January* 20*th*. A quantity of barley, soaked as in former experiments, was introduced into a jar filled with and inverted over mercury. At the expiration of 12 days a very considerable quantity of gas was produced, at least five or six times the bulk of the barley; but nothing like vegetation was perceivable. The gas on examination was found to consist of carbonic acid, being entirely absorbed by lime-water. The barley had not the least sweet taste, nor did it appear to have undergone any sensible change.'

This experiment, and other early ones, are perhaps not to be regarded as highly critical, especially in regard to the complete exclusion of oxygen and micro-organisms, but later observations showed, without a doubt, that the conclusion derivable from these early experiments was correct, and that an evolution of carbon dioxide by aerobic plants in absence of oxygen is a general phenomenon. This evolution of carbon dioxide in absence of oxygen was described as 'intramolecular' by Pflüger, who in 1875 observed the phenomenon in the frog, the idea involved in this term being that the carbon and oxygen of the exhaled carbon dioxide must come together within the molecules of the substance of the animal. This term was carried over into plant physiology by Pfeffer, but it is not a very happy one, and the expression 'anaerobic respiration' introduced by Kostychev in 1902 was for many years almost universally employed by writers in English, although 'intramolekulare Atmung' continued in use by some German writers.

In establishing the general existence of anaerobic respiration, and in confirming the earlier observation that along with the evolution of carbon dioxide, alcohol is produced, Pasteur and his pupils Lechartier and Bellamy played a prominent part eighty years ago. This discovery of alcohol as a product of anaerobic respiration at once suggested a comparison of the latter process with yeast fermentation in which the products are also carbon dioxide and alcohol. Although Sachs did not accept the view that there is a connexion between these two processes, and indeed considered 'that the formation of alcohol in the absence of oxygen is an abnormal process throughout, and has nothing to do with ordinary respiration,' the evidence since Pasteur's time for the similarity of anaerobic respiration and alcoholic fermentation has grown rather than diminished, as will appear from a consideration of the evidence presented in the next chapter.

As a result of this, there has been a tendency of late years, as mentioned in the last chapter, to use the word fermentation to denote the process otherwise known as anaerobic respiration. In favour of this is the fact that a breakdown of carbohydrate to carbon dioxide and alcohol does in certain circumstances occur in presence of oxygen, so that it is not very appropriate to describe the process as anaerobic. On the other hand, in spite of the weight of evidence in favour of the identity of anaerobic respiration with yeast fermentation, there are some instances in which ethyl alcohol does not appear among the end products of anaerobic breakdown, and many more in which the quantity of alcohol produced is much less than the theoretical amount that would be produced in fermentation. A more serious objection to the use of the term fermentation in place of anaerobic respiration is that some of the advocates of its use would not regard anaerobic breakdown as respiration, which term they would restrict to those breakdown processes depending on molecular oxygen. As Seifriz has pointed out, the older concept of respiration as the processes in living tissues whereby energy is liberated is the broader one and therefore to be preferred. Indeed, as regards the question of the amount of ethyl alcohol produced in anaerobic respiration, only exceptionally is the amount found that we should expect if the sugar is completely respired to carbon dioxide and alcohol. The equation representing this reaction, namely,

$$C_6H_{12}O_6 = 2C_2H_5OH + 2CO_2$$

indicates that equal molecular quantities of alcohol and carbon dioxide are produced. Since the molecular weights of these substances are respectively 46 and 44 the quantities should also be almost equal in weight ($CO_2/C_2H_5OH = 0.96$). An approximate equality of output of the two products was actually observed in 1901 with germinating seeds having carbohydrate reserves (*Pisum sativum* and *Vicia Faba*) by Godlewski and Polzeniusz, who obtained a ratio of CO_2/C_2H_5OH varying between 0.975 and 1.096, a relation confirmed by Nabokich in 1903, who obtained a value of 0.984. Subsequent observations, especially by Kostychev and Boysen Jensen, have, however, shown that this agreement with theory is by no means general, as an inspection of Table XII shows.

From this table it will be observed that among higher plants, only exceptionally is the amount of alcohol found in anaerobically respiring tissues equal to that of the carbon dioxide evolved. In potato tubers ethyl alcohol may be completely absent. Two explanations of this state of affairs are possible. Either the alcohol is produced in the amount indicated by the equation given above and is immediately utilized in some secondary reaction, or the respiration process does not always take the same course, and other products are formed besides, or instead of, ethyl alcohol. This question is further considered below.

The term zymasis used by Meirion Thomas to denote the breakdown of carbohydrate to carbon dioxide and ethyl alcohol, whether under aerobic or anaerobic conditions, has much to recommend it. It does not, of course, include breakdown processes in which the zymase complex of enzymes is not concerned and in which therefore ethyl alcohol is not a final product.

According to Palladin, the formation of alcohol in anaerobic respiration only takes place when the supply of carbohydrate is sufficient. If this is not so, carbon dioxide is produced by the breaking down of some other cell constituent, and some other product may result.

Godlewski and Polzeniusz stated that alcohol and carbon dioxide were not produced in absence of oxygen

TABLE XII
RATIO OF ALCOHOL TO CARBON DIOXIDE FORMED IN ANAEROBIC RESPIRATION

Species	Organ	C_2H_5OH/CO_2	Observer
Mucor racemosus		0·99	Kostychev
Aspergillus niger		0·92	,,
Psaliota campestris	fruit body	0·00	,,
Pisum sativum	cotyledons	0·65	Boysen Jensen
,, ,,	,,	0·81	,, ,,
Acer platanoides	leaves	0·58	Kostychev
Prunus padus	,,	0·51	,,
Syringa vulgaris	,,	0·56	,,
Tropaeolum majus	,,	0·45	Boysen Jensen
,, ,,	,,	0·24	,, ,,
,, ,,	,,	0·17	,, ,,
Daucus carota	root	1·02	Kostychev
,, ,,	,,	0·91	Boysen Jensen
,, ,,	,,	0·86	,, ,,
,, ,,	,,	0·72	,, ,,
Brassica rapa	,,	0·49	Kostychev
Lepidium sativum	seedlings	0·57	,,
Sinapis sp.	,,	0·60	Boysen Jensen
,, ,,	,,	0·32	,, ,,
Pyrus malus, var. *sinap* (sweet apples)	fruit	0·82	Kostychev
Pyrus malus, var. *Anton* (sour apples)	,,	0·42	,,
Citrus aurantium (orange)	,,	0·70	,,
Vitis vinifera (green grapes)	,,	0·86	Boysen Jensen
		0·74	
Vitis vinifera (blue grapes)	,,	0·95	,, ,,
,, ,,	,,	0·88	,, ,,
,, ,,	,,	0·81	,, ,,
,, ,,	,,	0·74	,, ,,
Solanum tuberosum, var. *Magnum bonum*	dormant tuber	0·07	Kostychev
Solanum tuberosum, var. *Magnum bonum*	sprouting tuber	0·00	,,
Solanum tuberosum, var.	tuber	0·07	Boysen Jensen
,, ,,	,,	0·02	,, ,,
,, ,,	,,	0·00	,, ,,
,, ,,	,,	0·00	,, ,,

by germinating seeds of *Ricinus* which contain fat as the chief food reserve. However, during the germination of this and other fat-storing seeds, *Helianthus annuus* and *Cucurbita pepo*, Leach and Dent found a not inconsiderable output of carbon dioxide under anaerobic conditions.

It is significant in this connexion that Kostychev and Afanassjewa found that *Aspergillus niger*, when grown on carbohydrate media under anaerobic conditions, gave a yield of alcohol which was within a few per cent. of the theoretical amount. On the other hand, when this mould was grown on a peptone medium, no alcohol was produced. This is apparently correlated with the fact that on the latter medium the fungus forms no zymase, the enzyme complex which, as is well known, brings about the splitting of sugar into alcohol and carbon dioxide, whereas on a medium containing sugar these enzymes are produced. It therefore seems likely that the process of anaerobic respiration of *Aspergillus niger* may vary according to whether the mould is grown on sugar or peptone media. There is thus a possibility of different types of anaerobic respiration. On the other hand, when, in anaerobically respiring tissue, the amount of ethyl alcohol produced is not equivalent to the quantity of carbon dioxide formed, the fact that *some* ethyl alcohol usually accumulates rather suggests that ethyl alcohol is first formed and then used in a secondary reaction. There is, however, the possibility that some intermediate product, instead of producing ethyl alcohol, may give rise to some other substance or substances. Such an intermediate product might be pyruvic acid or acetaldehyde, the presence of which have been demonstrated in anaerobically respiring tissues and which are now commonly regarded as precursors of the production of alcohol in anaerobic respiration (see Chapter IV).

Miss J. W. Phillips has obtained some information about the products of anaerobic respiration of rice and barley. The amounts of carbon dioxide, alcohol and acetaldehyde produced by barley seedlings kept for various lengths of time in an atmosphere of nitrogen were determined. During the first 3·5 hours under this condition the rate of production of alcohol + aldehyde was only about one-third the rate of production of carbon dioxide, but with continued anaerobiosis the relative rate of production of alcohol + aldehyde increased until after about 9 hours their rate of production was not much different from that of carbon dioxide. It is worth noting that acetaldehyde was a constant product along with ethyl alcohol, amounting to about one per cent. of the quantity of the latter produced.

Some of the excess of carbon dioxide over alcohol + aldehyde produced during the first few hours in nitrogen might be accounted for as a loss of bound carbon dioxide released by the action of an acid, if such were produced in anaerobic respiration. Miss Phillips found that some lactic acid was produced as well as some unidentified volatile and non-volatile acids, but the total amount of these would not have brought about the release of sufficient carbon dioxide to account for the whole of the excess of this over the alcohol + aldehyde, so that there still remained unexplained a difference between the amounts of carbon dioxide and of alcohol + aldehyde. Whether other tissues behave similarly to barley leaves remains to be seen.

In later work with barley roots the same worker measured the loss in carbohydrate and production of alcohol and carbon dioxide by tissue kept for four hours in a continuous stream of nitrogen. She found that the amount of alcohol produced agreed well with what would be expected if the carbohydrate lost has been converted to alcohol and carbon dioxide, but that the amount of the latter was much in excess of the theoretical quantity. It would thus appear that the extra carbon dioxide must have arisen from some source other than carbohydrate, such as proteins or organic acids.

It is generally supposed that the capacity of aerobic

plants to remain alive for a time while deprived of an oxygen supply is directly related to their power of anaerobic respiration which supplies a certain amount of energy. The energy so released is, however, small in comparison with that produced during normal respiration. Thus in aerobic respiration, for every molecule of sugar completely oxidized to carbon dioxide and water, 674 calories are released, whereas the energy released in the splitting of one molecule of hexose to carbon dioxide and alcohol is usually stated to be 25 to 28 calories. Or put in another way, since six molecules of carbon dioxide result from the complete oxidation of one molecule of hexose, and only two molecules of carbon dioxide from the fermentation of one molecule of hexose, for every molecule of carbon dioxide formed in aerobic respiration about 112 calories are released as compared with only about 13 calories per molecule of carbon dioxide released in anaerobic respiration. An inspection of Table XIV indicates that the rate of anaerobic respiration of a plant or organ is rarely as high as its rate of aerobic respiration, so that on exclusion of oxygen the energy produced in respiration falls usually to one-ninth or less of its previous value. Only in a few tissues including those of some ripening fruits, as so far observed, is the ratio of the energy released in anaerobic respiration to that released in aerobic respiration likely to be a little higher than this.

Varying external conditions appear to affect anaerobic respiration, as far as can be judged from published results, in very much the same way as they affect aerobic respiration. As regards temperature, the most reliable data concerning the influence of this factor on anaerobic respiration are probably those of Fernandes published in 1923 and dealing with germinating peas respiring in an atmosphere of hydrogen. Fernandes' results with seedlings 21 hours old are shown graphically in Fig. 6. A comparison with the results for seedlings respiring aerobically, shown in Fig. 4, at once reveals how similar are the temperature effects in aerobic and anaerobic respiration. It will be observed that up to 30° C. the rate of respiration remains practically constant with time, but that above this temperature a time factor very obviously

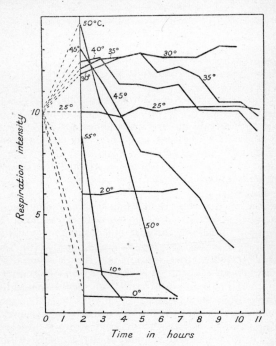

FIG. 6.—Curves showing the comparative rate of respiration of 21-hour-old seedlings of *Pisum sativum* as affected by different temperatures and related to time, when respiring in an atmosphere of hydrogen

(*After* Fernandes)

operates. The rates of respiration in air and hydrogen are compared in Table XIII.

These values give average temperature coefficients (Q_{10}) over the range of temperatures 0° to 30° of about

TABLE XIII

EFFECT OF TEMPERATURE ON GERMINATING PEAS 21 HOURS OLD IN AIR AND IN HYDROGEN

(*From* Fernandes)

Temperature	Respiration in hydrogen	Respiration in air
0	0·87	0·93
10	2·35	2·5
20	6·08	6·32
25	12·9	10·47
30	15·4–16·8	13·2–16·8

2·67 for respiration in hydrogen and about 2·52 for respiration in air. The values are of the same order of magnitude and the differences are almost certainly within the limits of experimental error. Probably no conclusion is justified from this similarity apart from a supposition that a purely chemical process determines the rate of respiration whether under aerobic or anaerobic conditions.

Some poisons have been found to affect aerobic and anaerobic respiration to about the same extent. Karlsen compared the aerobic and anaerobic respiration of wheat seedlings subjected to the action of various volatile poisons, namely, ether, benzene and ethyl alcohol. The relative effects of each poison upon the course of respiration in air and in nitrogen respectively were found to be similar. According to Marsh and Goddard, however, cyanides and carbon monoxide, which inhibit aerobic respiration of carrot root tissue, were found to have no effect on the anaerobic respiration of this tissue, while sodium azide, an inhibitor of aerobic respiration, had only a slightly depressing effect on anaerobic respiration.

Quite a number of attempts have been made to discover a quantitative relation between the intensities of aerobic and anaerobic respiration of the same organ. The first of these was made by Wortmann in 1880 on germinating seeds of *Vicia faba*; he found the rates of carbon dioxide evolution in air and in a Torricellian vacuum to be the same. Although subsequent measurements confirmed this result for this material, they show that there is in general no such agreement between the intensities of anaerobic and aerobic respiration. In Table XIV are shown the values published by Pfeffer in 1885 for the ratio of respiration intensity in hydrogen to respiration intensity in air $\left(\dfrac{R_h}{R_a}\right)$ for a number of plant organs.

From these values it will be observed that the intensity of anaerobic respiration is less than that of aerobic

TABLE XIV

RATIO OF ANAEROBIC TO AEROBIC RESPIRATION
(*From data published by* Pfeffer)

Species	Organ	$\dfrac{R_h}{R_a}$
Vicia faba	germinating seed	1·03 (mean of four determinations)
Triticum vulgare	seedling	0·49
Cucurbita pepo	,,	0·35
Sinapis alba	,,	0·18
Brassica napus	,,	0·25
Cannabis sativa	,,	0·34
Helianthus annuus	,,	0·33
Lupinus luteus	,,	0·24
Heracleum giganteum	unripe fruit	0·42
Abies excelsa	young leafy shoot	0·077
Orobanche ramosa	flowering stem	0·32
Arum maculatum	,, ,,	0·615
Ligustrum vulgare	leafy shoot	0·816
Lactarius piperatus	fruit body	0·32
Hydnum repandum	,, ,,	0·256
Cantharellus cibatius	,, ,,	0·666
Beer Yeast (on lactose)		0·310

respiration, as measured by rate of carbon dioxide output, in every case except that of germinating seeds of *Vicia faba*, while the actual ratio of the two intensities varies within wide limits. Subsequent determinations of the ratio by other observers have given similarly divergent results, the values varying between 0·177 and 0·181 recorded for seedlings of *Sinapis alba* by Pfeffer in 1885, and 1·3 and 1·1 found for green grapes by Boysen Jensen in 1923.

Boysen Jensen's work on this subject contains observations which suggest that the numbers found for the ratio of anerobic to aerobic respiration may not always have any definite value; for he shows that in two cases ex-

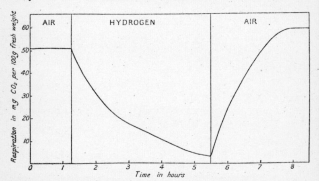

FIG. 7.—Graph illustrating the effect on carbon dioxide output of seedlings of *Sinapis alba* when changed from an atmosphere of air to one of hydrogen and *vice versa*

(*From data published by* Boysen Jensen)

amined, that of leaves of *Tropaeoleum majus* at 13° C. and seedlings of *Sinapis alba* at 16° C., the rate of respiration in hydrogen did not remain constant but fell off very definitely and considerably with time. This behaviour of the *Sinapis* seedlings is indicated graphically in Fig. 7, which shows that the average rate of respiration during the fourth hour in hydrogen was only about a quarter of that during the first hour. The behaviour of the *Tropaeoleum* leaves was very similar. It will be observed from Fig. 7 that in *Sinapis* seedlings, on replacing hydrogen by air the rate of respiration rises to above its original value. In the absence of information regarding the course of aerobic respiration during development, no conclusion can be drawn concerning this increased rate.

Considerable attention has been paid in recent years to the change in rate of carbon dioxide evolution by various tissues on transference from aerobic to anaerobic conditions. The first significant work on this question was that of Parija on apples. Parija's investigation formed the second of a series of *Analytic Studies of Respiration* carried out in F. F. Blackman's laboratory and published in 1928. The investigation of Blackman and Parija on the respiration of apples in air indicated that the fruit they used, belonging to the variety Bramley's Seedling, did not exhibit uniform behaviour in regard to the course of respiration, but it was concluded that the differences could be explained on the view that the apples belonged to three physiological classes which ripened at different rates. The respiration of these apples on transference to anaerobic conditions, in this case an atmosphere of pure nitrogen, was also dependent on the physiological class. With the slow-ripening apples, transference to nitrogen always brought about an immediate *increase* in the carbon dioxide output which rose to a maximum in a few hours and then fell rapidly to the level of respiration in air, at which level it continued for a shorter or longer time, finally declining rapidly and regularly. A typical example is shown in Fig. 8. With apples of the more rapidly maturing class (or classes) the rate of respiration rose immediately on substitution of nitrogen for air, and then declined slowly with fluctuations to a level well above that which would obtain in air. A typical example of this type is exhibited graphically in Fig. 9.

On re-transference to air the carbon dioxide output followed a fluctuating course, first rising, then falling and rising again, but ultimately reaching the normal rate of output for air, although sometimes not until the lapse of two days. Actually, on transference of the apple from air to nitrogen, the observed change in carbon dioxide output lagged behind the change in respiration intensity, owing to the time required for the respiratory carbon dioxide to diffuse from the tissues of the apple to the surrounding gas in the plant chamber. It is therefore reasonable to conclude that the actual rate of respiration in nitrogen

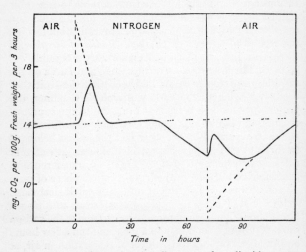

FIG. 8.—Graph illustrating the effect on carbon dioxide output of Bramley's Seedling apples belonging to the slow-ripening class, produced by change from an atmosphere of air to one of nitrogen, and *vice versa*

(*After* Blackman)

was highest at the moment of transfer to that gas and fell continually as shown by the broken lines in Figs. 8 and 9. This initial value of respiration rate in nitrogen was found to be about either 1·5 or 1·33 times the rate in air immediately before the replacement of air by nitrogen. As we shall see later, the value of this ratio was thought by Blackman to be of considerable importance in shedding light on the series of reactions involved in respiration.

ANAEROBIC RESPIRATION

This higher rate of carbon dioxide output in nitrogen as compared with the rate in air is by no means general. It has, however, been observed in carrot roots by Choudhury and it is possible that high rates of anaerobic respiration are characteristic of senescent fruits. Thus G. R. Hill in 1913 observed the rates of anaerobic respiration as measured by carbon dioxide output were

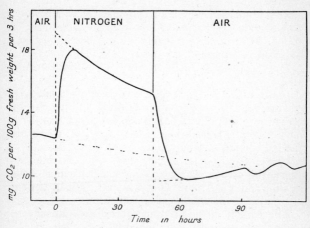

FIG. 9.—Graph showing effect of an atmosphere of nitrogen on the carbon dioxide output of Bramley's Seedling apples of the rapidly ripening class

(*After* Blackman)

as high, or higher, than those of aerobic respiration, in senescent grapes, cherries, and blackberries. It has already been mentioned that Boysen Jensen in 1923 obtained similar results for the first of these fruits. Immature fruit does not, as far as observations go, behave in this way. An important point connected with the above fact is that since three times as much sugar is required to produce a definite quantity of carbon dioxide anaerobically as is required to produce it aerobically, it follows that the destruction of sugar or other respiratory substrate must proceed three or more times as rapidly in absence of oxygen as in its presence.

Work on similar lines was carried out by Leach and Dent on the changes produced in the respiration of germinating seeds when they are successively subjected to atmospheres of air, nitrogen and again air. The results obtained show that germinating seeds when they are placed in nitrogen exhibit a respiratory behaviour which differs markedly from that observed by Blackman in apples. The species used were *Ricinus*, *Helianthus* and *Cucurbita* as representatives of fat-storing seeds and *Lathyrus*, *Zea* and *Fagopyrum* as carbohydrate-storing seeds.

In the fat-storing seeds the change from air to nitrogen produces an immediate and rapid fall in the rate of carbon dioxide output. This initial fall is followed by a more gradual fall in carbon dioxide production and this last fall continues throughout the anaerobic period.

The change from nitrogen to air produces an opposite effect, the carbon dioxide output increases first of all rapidly and then more slowly as the normal aerobic respiration of the seedlings is resumed.

The carbon dioxide output of carbohydrate-storing seeds shows a similar rapid initial fall when the change from air to nitrogen is made. It also shows a similar rapid rise and subsequent resumption of a normal course when the seedlings are transferred from nitrogen to air. The respiratory behaviour however of carbohydrate seeds in nitrogen during the period immediately following the initial fall in carbon dioxide production is peculiar. After this initial fall, the output of carbon dioxide first rises for a time and then again falls, this last fall being continued throughout the rest of the anaerobic period. It would appear from this that, with carbohydrate-storing seeds, the change from aerobic to anaerobic conditions brings into existence a fresh source of carbon dioxide, but as the substrate which produces this new carbon dioxide output is limited in quantity and is only produced in presence of oxygen, the rise in respiratory activity produced by it can be maintained only for a short time.

The final gradual fall in anaerobic carbon dioxide production shown by all the seeds used might further appear to indicate that in all cases the substrate for anaerobic respiration is dependent in the first place upon the presence of oxygen for its production.

Measurements of the respiratory quotients of the experimental seedlings were made and these have thrown some additional light on the effect of anaerobic conditions. With all the seedlings used, the respiratory quotient falls rapidly for a time when the change from nitrogen to air is made. This fall indicates a high rate of oxygen absorption by the seedlings, a fact which indicates that during the period in nitrogen some readily oxidizable substance accumulates in their cells. The fall in the quotient is followed by a rise which continues until the normal value for the particular seedling is reached.

The numerical relationships between the final respiration rate in air and the initial respiration rate in nitrogen when these various seedlings are transferred from air to nitrogen, are shown in Table XV and may be compared with the values obtained by Blackman for apples. (Cf. p. 72.)

TABLE XV

THE RELATIONSHIP BETWEEN RESPIRATION IN AIR (OR) AND RESPIRATION IN NITROGEN (NR) OF YOUNG SEEDLINGS

Seedling	$\frac{NR}{OR}$
Lathyrus odoratus	0·22
Fagopyrum esculentum	0·35
Zea mais	0·25
Helianthus annuus	0·52
Cucurbita pepo	0·40
Ricinus communis	0·51

An examination of the figures given in the above table show that with carbohydrate seeds (*Lathyrus*, *Fagopyrum* and *Zea*) a change from respiration in air to respiration in nitrogen results in a reduction in the carbon dioxide output to one-third or less than one-third. With the three fat-storing seeds the reduction is however not so great. In these, the relatively high values of $\frac{NR}{OR}$ may in part be due to an insufficient supply of oxygen, when the seedlings are respiring in air, to allow OR to attain its full value. A suggested reason for this is that oxygen, besides being used for respiration, is in fat-storing seeds used for the conversion of fat into carbohydrates.

It has already been pointed out that where the substrate is sugar a ratio of $\frac{NR}{OR}$ exceeding one-third indicates that the rate of sugar destruction under anaerobic conditions must exceed that under aerobic conditions, assuming that the products of aerobic respiration are carbon dioxide and water, and those of anaerobic respiration carbon dioxide and ethyl alcohol. The presence of oxygen would then appear to lessen the rate of sugar destruction. This is known as the Pasteur effect after the distinguished investigator who first recorded it. As well as in senescent fruits it has been observed in a number of other tissues, including those of various storage organs such as carrot root, as recorded by Choudhury, Turner and Marsh and Goddard, roots of red beet root and tubers of Jerusalem artichoke as described by Choudhury and by Stiles and Dent, and parsnip roots as observed by Appleman and Brown. The significance of the Pasteur effect will be discussed in the next chapter.

Reference has been made in the preceding chapter to the effect of oxygen concentration on respiration. It was there shown that when the oxygen concentration is reduced below a certain level the respiratory quotient increases progressively with decrease in the oxygen concentration. This is reasonably explained on the assumption that in these low oxygen concentrations the carbon dioxide is produced partly by the process of aerobic respiration, and partly by the process of anaerobic respiration, or, as some recent writers would prefer to put it, fermentation. As the oxygen concentration is lowered the proportion of the latter increases until in complete absence of oxygen the whole of the carbon dioxide is produced anaerobically. Put in another way, with increase in oxygen concentration the anaerobic component becomes less and less a fraction of the whole until when a certain oxygen concentration is reached it is suppressed altogether. This oxygen concentration was called by Meirion Thomas the extinction point. The numbers given in Tables V and VI suggest that it may often be in the neighbourhood of 5 per cent.

It has been suggested that anaerobic respiration may normally take place where diffusion of gases between the tissues and outer air is slow. Such a state of affairs may prevail in germinating seeds provided with testas of a low degree of permeability, in middle parts of bulky tissues such as large fruits, storage roots and tubers and various organs of succulent plants, in woody stems surrounded by a cork layer, and in organs submerged in water. In all such organs it is at any rate a possibility that the actual oxygen concentration in the respiring cells may be low, while in addition there may be a partial accumulation of carbon dioxide so that the concentration of this is maintained at an unusually high level and so may retard respiratory activity. (Cf. p. 51.)

As regards bulky tissues, there is evidence that respiration may indeed lead to high concentrations of carbon dioxide and low concentrations of oxygen. Boswell and Whiting in 1940 found that in potato tubers weighing about 70 gm. the average concentration of carbon dioxide in the internal atmosphere was about 11·4 per cent. The still higher value of 34·1 per cent. was found twenty years earlier by Magness for potato tubers at 22° C., while the same investigator recorded a concentration of 28·6 per cent. carbon dioxide in carrot roots at 24° C. The corresponding concentrations of oxygen in these two organs were 5·7 and 5·2 per cent. respectively. As these values were the average for whole cylinders cut out with a cork borer it may be concluded that the actual concentrations in the middle regions of the organs were even further removed from those of carbon dioxide and oxygen in normal air. Reference has already been made to the occurrence under certain conditions of high concentrations of carbon dioxide in the internal atmospheres of Gladiolus corms, as recorded by Thornton and Denny (cf. p. 53).

On the other hand, Devaux found in 1891 by direct analysis that the internal atmosphere of large cucurbitaceous fruits, as, for example, those of *Cucurbita maxima* and *C. melanosperma*, contained nearly as high a concentration of oxygen as is present in atmospheric air, percentages of this gas of 18·29 and 17·89 being found in these two species, respectively. Meirion Thomas found only very small quantities of ethyl alcohol and acetaldehyde in freshly gathered apples, namely 0·006 per cent. by weight of the former and 0·0005 per cent. by weight of the latter. These observed facts suggest that, as far as they go, very little anaerobic respiration is likely to occur in fleshy fruits. This does not necessarily mean that the aerating system is adequate to allow of rapid enough diffusion of gases for oxygen respiration to proceed normally. The negligible amount of anaerobic respiration might result from a high concentration of carbon dioxide in the interior of the fruit depressing the rate of anaerobic respiration.

However, the production of carbon dioxide and alcohol and acetaldehyde in large fruits would appear to present a special problem. As long ago as 1896 Gerber had suggested that anaerobic respiration always takes place in these and that through this process arise the alcohols, aldehydes and esters which are normally present in such fruits. The investigations of Meirion Thomas indicate that the production of appreciable amounts of alcohol and acetaldehyde in apples does not result from anaerobic conditions in the interior of these fruits but from changes in the system or systems concerned in the breakdown of carbohydrate. As already noted, only negligible quantities of alcohol and aldehyde were found in freshly gathered apples respiring in air. Kept in an atmosphere of nitrogen, however, similar apples were found by Thomas to accumulate considerable quantities of alcohol, the percentage of this rising to 0·17, 0·25 and 0·39 after 15, 23 and 38 days respectively at 1° C. Thomas therefore concluded that in freshly gathered healthy apples in air no significant amount of anaerobic respiration, or zymasis as he called it, occurred, but that it does as usual under anaerobic conditions.

As apples become senescent, however, the behaviour of the fruit changes. Thomas and Fidler examined the production of carbon dioxide and alcohol and acetaldehyde in apples at different stages in their development and storage exposed to mixtures of nitrogen and oxygen at 23° C. They found, as would be expected, that with increasing proportions of oxygen in the gas the amount of anaerobic respiration or zymasis was reduced until at the extinction point no alcohol was formed. For apples still on the tree or early in the storage season the extinction point appeared to be not higher than 2·5 per cent. for apples of the varieties Newton Wonder and Bramley's Seedling, but as the season advanced progressively higher concentrations of oxygen were required to suppress zymasis, so that in apples that had been long in storage zymasis might even occur in 100 per cent. oxygen.

Thomas found that zymasis also occurred in apples under aerobic conditions in the presence of high concentrations of carbon dioxide. This CO_2-zymasis was found to differ from anaerobic zymasis in that during the former process more acetaldehyde accumulated than during anaerobic zymasis. Later Thomas and Fidler found that hydrocyanic acid could also induce zymasis in apples under aerobic conditions. Here also the proportion of acetaldehyde to ethyl alcohol produced was higher than in anaerobic zymasis. These results are important in their bearing on the mechanism of respiration, and further reference will be made to them in this connexion in the following chapter.

The aerobic production of ethyl alcohol has also been observed by Gustafson in tomatoes during ripening, the percentage increasing from 0·0011 in small green fruits 1 to 2 cm. in diameter to about 0·014 in red-ripe fruits. Here also the production of ethyl alcohol increased greatly during a period in nitrogen, and, as in apples, a small amount of acetaldehyde was found along with the alcohol.

CHAPTER IV

THE MECHANISM OF RESPIRATION

THE CONNEXION BETWEEN FERMENTATION, ANAEROBIC RESPIRATION AND AEROBIC RESPIRATION

WE have seen that normal aerobic respiration consists broadly of the oxidation of carbohydrates or other organic material into carbon dioxide and water. Since this means the breaking down of a complex molecule containing at least six carbon atoms into carbon dioxide with but one, it is extremely unlikely that the respiratory process takes place in a single step. One of the aspects of an inquiry into the mechanism of the respiratory process is, therefore, the determination of the probable stages in the breaking down of the carbohydrate or other complex material utilized in respiration.

Further, we are well aware that under temperature conditions similar to those prevailing in a living organism no breaking down of carbohydrate into carbon dioxide and water takes place if we merely supply carbohydrate with oxygen. A second aspect of the inquiry into the mechanism of respiration is therefore concerned with an examination of the special conditions of the living cell which enable this catabolism to take place. Here obviously we have to consider the various oxidation systems, enzymatic and otherwise, which are known to be present in the cell, as well as enzymes which split off carbon dioxide from more complex substances, and to decide, as far as we are able, what connexion, if any, they may have with the respiratory process.

These two aspects of our problem are, of course, intimately connected, for, from what we know of the characteristics of enzyme actions, it is at least a possibility that every stage in the process is catalysed by its own enzyme. It will therefore not always be possible to separate them in discussion.

In attempting to formulate a theory of the course of respiration in plant cells much use has been made of information obtained from investigations on yeast fermentation. The value of this information in regard to the problem of normal aerobic respiration depends largely on (*a*) to what extent fermentation by yeast is identical with anaerobic respiration, and (*b*) the connexion between anaerobic and aerobic respiration. With regard to the former of these questions it has often been assumed that yeast fermentation and anaerobic respiration are the same process. There are several facts in favour of this view. In fermentation, hexose sugar is utilized and the same is frequently the case in anaerobic respiration; where some other substance is utilized it is possible, and even probable, that the substance is first transformed into hexose sugar before it can be utilized for respiration. In both processes carbon dioxide is produced, while in many instances of anaerobic respiration the formation of ethyl alcohol has been demonstrated. Where this has not been done, and where the quantity of alcohol formed, relative to the carbon dioxide produced, is less than that formed in fermentation, the absence or shortage of this substance can be attributed to its utilization in some secondary reaction. Thus, as far as is known, both the substrate and products in anaerobic respiration are frequently the same as those in alcoholic fermentation.

Of course, it does not follow that because the substrate and end products are the same, that these latter have been produced by the same mechanism. There are, however, additional pieces of evidence suggesting that this is indeed so. The enzyme, or rather group of enzymes, known as zymase, which was already known to be the catalyst bringing about the breaking down of hexose to alcohol and carbon dioxide in yeast fermentation, was shown to be present in the cells of a number of higher plants by Stoklasa and Czerny in 1903. Since then there have been isolated from yeast a number of enzymes which in all probability form part of the zymase complex. Further reference will be made to these later. It will suffice here to mention that several of these enzymes which play their respective parts in the various stages of yeast fermentation have also been isolated from various higher plants.

Again, Kostychev showed that acetaldehyde was probably formed as an intermediate product in alcoholic fermentation, and Neuberg showed that by the addition of a sulphite to the fermenting liquor, the aldehyde could be fixed as aldehyde sulphite and so caused to accumulate. Neuberg and Cohen similarly succeeded in fixing aldehyde in the anaerobic respiration of various fungi, while in the same way Neuberg and Gottschalk fixed aldehyde formed in the anaerobic respiration of pea seeds. Previous to this, in 1913, Kostychev, Hübbenet, and Scheloumov had demonstrated the formation of acetaldehyde in anaerobically respiring poplar flowers.

One other point of resemblance between yeast fermentation and anaerobic respiration is the effect of the addition of phosphate. Fermentation of sugar by means of expressed yeast juice or dried yeast is accelerated by the addition of soluble phosphate, while addition of phosphate to the tissues of higher plants killed in various ways also leads to an increased rate of carbon dioxide production. Although Kostychev held that where phosphate brought about an increased rate of carbon dioxide output from dead tissues this was merely due to the effect of the phosphate in increasing the alkalinity of the medium, the school represented in particular by Zaleski, L. Ivanov and N. Ivanov some forty years ago held that phosphate did actually accelerate respiration. In 1924 Lyon found that phosphate accelerated both aerobic and anaerobic respiration.

The evidence in favour of the view that alcoholic fermentation is the same process as anaerobic respiration is thus very strong. We are justified, in the absence of any definite evidence to the contrary, in assuming, at least provisionally, that fermentation of sugar by yeast and anaerobic respiration follow the same course.

The relationship between aerobic and anaerobic respiration is not so clear. That a close connexion exists between these processes was first suggested by Pflüger in respect of certain animals which could carry on respiration anaerobically for a time. His idea was that the respirable material was first broken up anaerobically into carbon dioxide and easily oxidizable products, the latter then being oxidized by atmospheric oxygen to carbon dioxide and water.

Although such authorities as Nägeli and Sachs held the evolution of carbon dioxide in absence of oxygen to be a pathological phenomenon due to injury resulting from absence of oxygen, and therefore to have no connexion with normal respiration, the view of Pflüger was taken over by Pfeffer into plant physiology. He suggested that ordinary aerobic respiration takes place in two stages, the first, a splitting of sugar by a number of steps into alcohol and carbon dioxide, the second, the oxidation by atmospheric oxygen of the alcohol, or some other product formed in one of the steps of anaerobic respiration, into carbon dioxide and water. The first of these stages is independent of oxygen and is in fact the process called anaerobic respiration.

This theory found little support at the time, and Pfeffer himself later appeared to give it up. Its lack of support was largely due to two causes. In the first place no constancy was found in the ratio between the intensities of anaerobic and aerobic respiration for different plant organs. In the second place a series of experiments by Diakanov demonstrated that anaerobic respiration of certain fungi could only take place on saccharine media, although aerobic respiration took place in a wide range of other media.

THE MECHANISM OF RESPIRATION

With regard to the first of these difficulties in the way of accepting Pfeffer's hypothesis it was argued that if anaerobic respiration is the first stage of normal respiration, the intensities of the two processes should bear a constant relation to one another. The numbers given in Table XIV, Chapter III, indicate that this is not so, and that the ratio of the respiration rate in hydrogen to that in air may vary widely from plant to plant and organ to organ. This difficulty in the way of accepting the Pfeffer theory is, however, only apparent. If the theory were correct there is no reason why the ratio of the intensities of anaerobic respiration and aerobic respiration should be constant. In the first place, under anaerobic conditions the rate of respiration will depend, at the beginning, on the concentration of the sugar; later, if alcohol accumulates, the latter will tend to retard the process more and more, not merely in accordance with the law of mass action, but because of its toxic effect on the respiring cells. We have already noted that it is highly probable that alcohol accumulates to different degrees in different plant organs. We may therefore add that it seems likely that the rate of anaerobic respiration in different plants and different plant organs is a function of the amount of alcohol accumulated at any time, and that this varies very considerably from plant to plant. Under aerobic conditions, provided the resulting alcohol is oxidized as soon as it is formed (and the absence of alcohol in the tissues would support this view), the rate of aerobic respiration will suffer no retardation on account of accumulation of products.

The second difficulty, which was the outcome of Diakanov's experiments, was proved by Kostychev to be also without foundation. Kostychev repeated Diakanov's work and showed that his experimental results were largely vitiated by the toxic action of the products of metabolism under anaerobic conditions. When this toxic action was eliminated no distinction could be found between saccharine and non-saccharine nutrients such as glycerol, mannitol, and lactic acid when used as substrates for anaerobic respiration.

Thus these particular objections to the theory of a close connexion between anaerobic and aerobic respiration disappear. Further, although there is no direct evidence in favour of such a connexion, there are several facts which indirectly suggest that it exists. These are as follows:

1. Anaerobic respiration of normally aerobic plants when deprived of oxygen, appears to be a universal phenomenon. It is true that Lyon failed to observe the evolution of carbon dioxide from *Elodea* in absence of oxygen, but this appears to be an isolated and exceptional case.

2. The enzyme complex zymase, which, as we know, is concerned in the anaerobic splitting off of carbon dioxide from sugar, appears to be universally present in plant cells. It would thus appear that the process in which zymase is concerned is part of the normal respiratory mechanism of the cell. Since the action of zymase is not suppressed by oxygen, the absence of alcohol production in presence of oxygen cannot be explained as due to the inhibition of zymase.

3. If the first stage (or stages) of normal respiration consists of the anaerobic production of easily oxidizable substances, a period of anaerobic respiration should lead to the accumulation of such substances. Subsequent transference to aerobic conditions should then, owing to the increase in concentration of the substrate for the oxidation process, result in a rate of respiration above the normal. Such an increased rate of respiration after a period of anerobiosis was observed by Maquenne in 1894 and has subsequently been recorded by Palladin and other observers. Where such an increase is not recognizable it may sometimes be accounted for by the toxic effect of the products of anaerobiosis.

4. It has been put forward by Kostychev that the essential plant-oxidizing systems, to which reference will be made shortly, are incapable of oxidizing sugars, but

are able to oxidize substances present in fermented sugar solutions. If the first stages of aerobic and anaerobic respiration are the same it is readily understandable that the substance actually oxidized in the process of aerobic respiration is not sugar but a decomposition product of sugar resulting from the action of enzymes of the zymase complex.

5. It has been mentioned that the acetaldehyde is in all probability an intermediate product in anaerobic respiration. It is therefore significant that Klein and Pirschle demonstrated the formation of acetaldehyde during normal aerobic respiration. Further, there is every reason to suppose that in fermentation and anaerobic respiration the acetaldehyde is produced by the action of carboxylase, one of the enzymes of the zymase complex, on pyruvic acid. James and Norval have demonstrated the similar formation of acetaldehyde from pyruvic acid in barley tissues.

There is thus a considerable weight of evidence in favour of the theory of a close connexion between anaerobic and aerobic respiration. However, the simple view which considers anaerobic respiration to be the first stage of aerobic respiration is unlikely to be correct, for ethyl alcohol is even less easily oxidized than hexose sugars. Pfeffer's alternative theory has been urged by Kostychev and meets this difficulty. According to this theory, anaerobic respiration or alcoholic fermentation itself takes place in several stages, a formation of labile intermediate substances produced by the action of zymase preceding the production of alcohol. Under anaerobic conditions, these intermediate substances pass over into alcohol, but when oxygen is present the labile substances are oxidized to carbon dioxide and water. From what has been stated previously it would appear that acetaldehyde might be such an intermediate substance yielding alcohol or becoming completely oxidized according to conditions. This theory may be represented schematically thus:

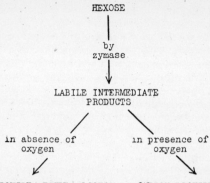

Current views of the mechanism of respiration may be regarded as an elaboration of this theory, although they have actually developed more or less independently of it.

Not all plant physiologists have accepted the theory of the close connexion of aerobic and anaerobic respiration. Thus in 1923 Boysen Jensen pointed out that in certain plant material, including *Tropaeolum* leaves, *Sinapis* seedlings, *Aspergillus niger* and *Penicillium glaucum*, the ratio of the rate of anaerobic to the rate of aerobic respiration sinks below 1/3 without the material suffering injury. Now since one molecule of hexose sugar yields six molecules of carbon dioxide by complete oxidation, and two molecules of carbon dioxide by fermentation or anaerobic respiration, it follows that in such cases anaerobic processes do not split up enough sugar to account for the whole of the aerobic respiration. Subsequently the same author pointed out that D. Müller had prepared an enzyme, glucose oxidase, from *Aspergillus niger*, which could oxidize glucose to gluconic acid directly without the intervention of zymase, so that breaking down of hexose by the latter was not a necessary preliminary for oxidation. Further, E. Lundsgaard found that zymase was easily paralysed by monoiodoacetic acid, so that baker's yeast treated with this substance in suitable concentration had very little fermenting power; it could, however, oxidize sugar to carbon dioxide and water, so that again zymase was not necessary in the chain of reactions involved in the complete breaking down of sugar. It was found that glucose oxidase was rather resistant to the action of monoiodoacetic acid.

From a consideration of these facts, Boysen Jensen concluded that some organisms are able to oxidize sugar directly without its being first subjected to splitting by zymase.

In 1938 Turner published the results of an investigation on the effects of sodium monoiodoacetate on both the aerobic and anaerobic respiration of carrot root tissue. He found that both processes were inhibited by the reagent in the same way, but that aerobic respiration was affected less than anaerobic respiration. Turner supposed that this effect might result from oxygen reducing the inhibitory influence of the iodoacetate, and that his findings were not out of harmony with the Pfeffer-Kostychev theory of a connexion between aerobic and anaerobic respiration.

It may therefore be concluded (1) that where the final products of anaerobic respiration are ethyl alcohol and carbon dioxide there are reasonable grounds for regarding the process as the same as fermentation by yeast, and (2) that the breakdown of carbohydrate in aerobic respiration and fermentation follows the same course up to a certain stage, but that the fate of the intermediate products at this stage depends on the presence or absence of oxygen, the final products being carbon dioxide and ethyl alcohol in absence of oxygen, and carbon dioxide and water in its presence. In low oxygen concentrations both kinds of breakdown may occur together.

THE RESPIRATORY SUBSTRATE

Before discussing in detail the course of fermentation and respiration a consideration of the substrate in these processes is desirable. Already in the first two chapters it has been indicated that these may be carbohydrates, fats and, in certain conditions, proteins. Carbohydrates which may be drawn on in different plants include not only hexose sugars, chiefly glucose and fructose, but various more complex substances such as disaccharides, particularly sucrose, and polysaccharides including starch, inulin and the so-called hemicelluloses or reserve celluloses which are frequently not celluloses at all but condensation products of other sugars such as galactose and mannose. Probably glycosides, compounds of sugar with other groupings, may also provide material for respiration. With many of these more complex carbohydrates it seems probable that as a first step in their utilization they are hydrolysed by the appropriate enzyme system to the hexose sugar level. Thus sucrose, by means of the enzyme sucrase, would be converted to equal quantities of glucose and fructose, inulin by the action of inulase to fructose, starch by the action of amylase and maltase to glucose. In the utilization of starch, at any rate, it is possible that glucose itself is not necessarily produced, for starch is broken down not only by amylase but by the enzyme phosphorylase which in presence of a phosphate effects the breakdown of starch with the production of a compound of glucose and phosphoric acid, glucose-1-phosphate, as the final product of the action, and this substance, as will appear later, can be utilized directly in fermentation.

Where fats are the substances utilized in respiration it is generally supposed that hexose sugar is formed from them and that this can be regarded as the actual substrate. Analyses of germinating seeds and of seedlings make it perfectly clear that there is such a conversion of fat to sugar although the mechanism of this change is not understood. It may be presumed that the fats are first hydrolysed to fatty acids and glycerol by the action of the enzyme lipase and that these are subsequently converted into hexose. But it is possible that material derived from fat and utilized in respiration may not go through a hexose stage, for, as will be shown later, phosphorylated derivatives of glycerol appear as substances formed during the breakdown of hexose in fermentation.

The mode of degradation of proteins when these are used in respiration is even more obscure. It has been pointed out earlier that the end products of protein respiration are carbon dioxide, water and asparagine or, with complete degradation of the substrate, carbon dioxide, water and ammonia. It would seem probable that the first stage in the utilization of protein would be the action of protease enzymes in splitting up the protein to its constituent amino-acids. The changes by which asparagine and ammonia are then produced are by no means clear. Yemm has pointed out that there is evidence that the aerobic breakdown of amino-acids begins with the removal of the $—NH_2$ group with formation of the corresponding α-ketonic acid:

$$R.CH(NH_2).COOH + O = R.CO.COOH + NH_3$$

The action of the enzyme carboxylase, one of the earliest constituents of the zymase complex to be recognized, would then bring about the production of the corresponding aldehyde and carbon dioxide:

$$R.CO.COOH = R.CHO + CO_2$$

The production of aspargine presumably results from the amination of acids containing four carbon atoms, but this probably involves a series of changes rather than a direct amination of aspartic acid produced immediately from protein hydrolysis as only a comparatively small quantity of this amino acid is usually present.

However, whatever may be the course of protein degradation it is at least possible that this may differ over much of its path from that of normal aerobic respiration of hexose sugar.

In fermentation and respiration, therefore, although we may regard the usual substrate as hexose sugar, we must be prepared to recognize that reserve substances may give rise to compounds other than hexoses which are utilized in fermentation or respiration.

It will be desirable at this point to consider the constitution of the sugars. While a number of different structural formulae were ascribed to the various sugars in the past, the views of Haworth are now practically universally accepted. According to him, the molecule of ordinary glucose and of other sugars of the aldose series is best represented as a six-atom ring comprising five carbon atoms and one oxygen atom, the sixth carbon atom forming part of a —CH$_2$OH group which constitutes a side chain to the ring. Ordinary glucose exists in two stereo-isomeric forms, α and β, of which the latter may be represented by the formula

$$\begin{array}{c}\text{[β-glucose ring structure]}\end{array}$$

In this sugar the hydrogen atoms are regarded as lying alternately above and below the plane of the ring. In α-glucose the position of the hydrogen atom and the hydroxyl group attached to one of the carbon atoms is reversed so that this steroisomer is represented by the formula

$$\begin{array}{c}\text{[α-glucose ring structure]}\end{array}$$

These two isomers are both obtainable, and in solution can pass by mutarotation from one to the other, while both give stable derivatives such as the well known α- and β-methyl glycosides.

Sugars possessing such a formula can be regarded as derived from pyran

$$\begin{array}{c} O \\ CH \diagup \diagdown CH \\ \| \quad \| \\ CH \quad CH \\ \diagdown \diagup \\ CH_2 \end{array}$$

and are therefore termed pyranose sugars.

There are, however, a number of derivatives of glucose in which sugar appears to be present in a different form, that is, the atoms appear to be in some other arrangement. The first of these compounds to be recognized was γ-methyl glycoside and subsequently a number of other derivatives of glucose have been obtained in which the sugar part of the molecule has the same structure. Moreover, α and β forms of some, at least, of these derivatives of glucose have been isolated. Their chemical behaviour is such as to indicate that they are built up from a sugar having a five-atom ring and to which the name γ-glucose is given. The formula ascribed to this sugar is

$$\begin{array}{c} CHOH \\ \diagup \diagdown \\ HCOH \quad O \\ | \\ HCOH \text{——} C \text{——} CHOH \\ | \quad | \\ H \quad CH_2OH \end{array}$$

This sugar does not appear to have an independent existence, but since it enters into combination it is clear that the stable forms of glucose must be convertible into it under certain conditions, and that it can be regarded as a labile form of ordinary glucose. Such labile or

γ-sugars are said to belong to the furanose series on account of their relation to furan

$$\begin{array}{c} \mathrm{O} \\ \mathrm{CH} \quad \mathrm{CH} \\ \| \qquad \| \\ \mathrm{CH}\!-\!\mathrm{CH} \end{array}$$

Not only do sugars of the aldose series exhibit γ-forms, but those of the ketose series do also. Of the ketoses, one in particular, fructose, is almost universally present in plants, free or in combination with glucose as cane sugar. The formula of ordinary free fructose can be written as

$$\begin{array}{c} \mathrm{O} \\ \diagup \quad \diagdown \\ \mathrm{HO.CH_2.COH} \quad \mathrm{CH_2} \\ | \qquad\qquad | \\ \mathrm{CHOH} \quad \mathrm{CHOH} \\ \diagdown \quad \diagup \\ \mathrm{CHOH} \end{array}$$

and of γ-fructose as

$$\begin{array}{c} \mathrm{O} \\ \diagup \quad \diagdown \\ \mathrm{HO.H_2C.COH} \quad \mathrm{CH.CH_2OH} \\ | \qquad\qquad\qquad | \\ \mathrm{CHOH}\!-\!\mathrm{CHOH} \end{array}$$

The γ-form of fructose, like that of glucose, has never been obtained free, but it is in this condition that fructose occurs in combination in plants, for in both inulin and sucrose the fructose part of the molecule is in the γ-form. The free sugar on the other hand is in the normal six-atom (pyranose) ring form.

Without going further into this question, it may be pointed out how closely related are the various γ-sugars, including not only those of the hexoses, but also of the pentoses. An inspection of the following formulae will make the relation clear.

```
         O
        / \
    CH     CH
    ||     ||
    CH─────CH
       furan
```

```
           O
          / \
    CHOH     CH.CH₂OH
     |        |
    CH(OH)───CHOH
       γ-pentose
      (e.g. γ-xylose)
```

```
           O
          / \
    CHOH     CH.CH(OH)CH₂OH
     |        |
    CH(OH)───CHOH
     γ-aldo-hexose
    (e.g. γ-glucose)
```

```
             O
            / \
   HOCH₂COH     CH.CH₂OH
      |          |
      CH(OH)────CHOH
      γ-keto-hexose
     (e.g. γ-fructose)
```

It seems, therefore, that a conversion of γ-fructose into γ-glucose or γ-pentose and vice versa in reactions in the plant involving hexose is not impossible or even improbable. It is well known that the pyranose forms of glucose and fructose readily pass from one to the other in alkaline solution. Further, it is clear that normal glucose is transformed to the labile γ-form in the production of a number of glucose derivatives, as, for example, in the formation of γ-methyl glycoside from methyl alcohol and normal glucose, while not only can γ-fructose derivatives be produced from normal fructose, but the γ-fructose of sucrose is transformed to normal fructose (fructo-pyranose) on the inversion of the latter. Without definite evidence on the point, there is ground for supposing that a conversion in the plant of stable glucose and fructose into the labile and more highly reactive forms of these substances is likely.

Before leaving the question of the constitution of the hexoses it should be pointed out that it is convenient to number the carbon atoms in the sugar molecule. Thus for α-glucose (gluco-pyranose) the carbon atoms are numbered thus:

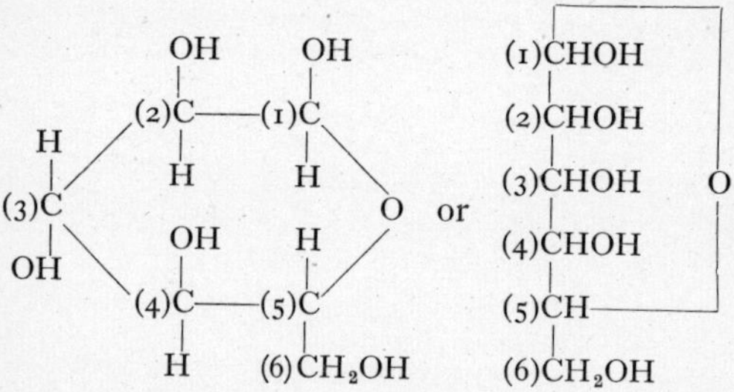

$$
\begin{array}{c}
\text{structure with numbered carbons (1)–(6) in pyranose ring}
\end{array}
\quad \text{or} \quad
\begin{array}{l}
(1)CHOH \\
(2)CHOH \\
(3)CHOH \\
(4)CHOH \\
(5)CH\!-\!\!-\! \\
(6)CH_2OH
\end{array}\!\!\Bigg]O
$$

and for γ-fructose (fructo-furanose) thus:

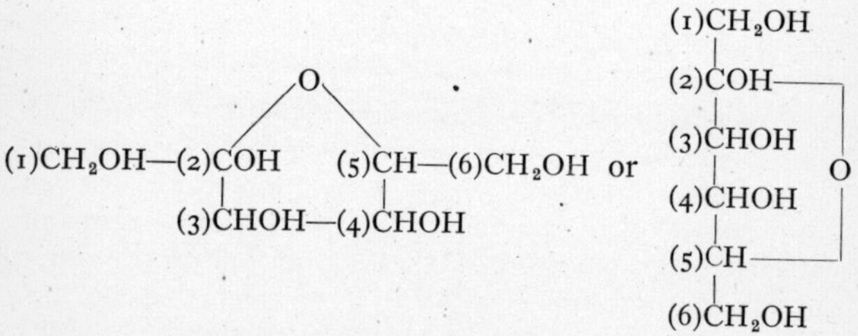

$$(1)CH_2OH\!-\!(2)COH \quad (5)CH\!-\!(6)CH_2OH$$
$$(3)CHOH\!-\!(4)CHOH$$

or

$$
\begin{array}{l}
(1)CH_2OH \\
(2)COH\!-\!\!-\! \\
(3)CHOH \\
(4)CHOH \\
(5)CH\!-\!\!-\! \\
(6)CH_2OH
\end{array}\!\!\Bigg]O
$$

The usefulness of this numbering will be obvious by reference to glucose phosphates which are formed during fermentation, in one of which the linkage of the phosphate grouping is through the hydroxyl attached to the first carbon atom, and the other in which the attachment is through the hydroxyl attached to the sixth carbon atom. It is simple and precise to refer to these two phosphates, whose formulae are

```
        O
        ‖  ⁄OH
(1)CH—O—P
        ⁀OH
 |
(2)CHOH
 |
(3)CHOH
 |
(4)CHOH
 |
(5)CH─────────┐
 |            O
(6)CH₂OH
```
and
```
(1)CHOH
 |
(2)CHOH
 |
(3)CHOH       O
 |
(4)CHOH
 |
(5)CH─────────┐
 |            |
              O
              |
         O    |
         ‖  ⁄OH
(6)CH₂—O—P
         ⁀OH
```

as glucose-1-phosphate and glucose-6-phosphate respectively. Similarly, the hexose diphosphate isolated by Harden and Young during fermentation (see p. 99) is fructose-1,6-diphosphate and if, as has been supposed, the fructose is in the furanose form its structure is represented by the formula

```
            O
            ‖  ⁄OH
(1)CH₂—O—P
            ⁀OH
 |
(2)COH─────────┐
 |             |
(3)CHOH        |
 |             O
(4)CHOH        |
 |             |
(5)CH──────────┘
 |
         O
         ‖  ⁄OH
(6)CH₂—O—P
         ⁀OH
```

THE COURSE OF FERMENTATION AND ANAEROBIC RESPIRATION

We are now in a position to consider the changes which occur when hexose sugar is broken down to carbon dioxide and ethyl alcohol as in alcoholic fermentation.

From what has already been written it will appear that these stages may be identical with those occurring in the anaerobic respiration of carbohydrate, at least when the end products are the same, and that up to a point the changes may be the same as those occurring in normal aerobic respiration. Since very much more definite evidence has been obtained of the course of fermentation than of respiration it will be convenient to deal with the former first and then relate the findings in regard to this to aerobic respiration. At the same time it should be realized that the identity of the first stages in respiration with those of fermentation is still a theory and not an established fact, albeit a theory with strong evidence in its support.

Our knowledge of the mechanism of the fermentation process really begins with the work of Buchner, who in 1897 obtained from yeast a product which brought about fermentation of sugar into alcohol and carbon dioxide without the intervention of the living cell. To this product, which possessed the general properties of an enzyme, the name zymase was given.

It was not very long before it became clear that zymase was not a single substance. In 1904 Harden and Young showed that preparations of zymase could be separated into three constituents, namely, (1) a colloidal part, the apoenzyme, (2) an organic crystalloidal part, the coenzyme, and (3) phosphate. The coenzyme was called cozymase by Euler and Myrbäck and is now more generally known as coenzyme 1. It is a complex substance the molecule of which is built up from a molecule of a purine called adenine, a molecule of the amide of nicotinic acid, two molecules of a pentose sugar d-ribose, and two molecules of phosphoric acid. It is diphosphopyridine nucleotide, and is now often represented by the symbol DPN. Subsequent work has shown that not only is there at least one other coenzyme in the zymase complex, but that the enzyme fraction contains a considerable number of different enzymes each responsible for one particular stage of the fermentation process.

As regards the phosphate fraction, not only did

Harden and Young find that fermentation by dried yeast or yeast juice was accelerated by phosphate, they found that in the fermentation of hexose, with the disappearance of the phosphate there was formed a compound of hexose and phosphate, namely, hexosediphosphate $C_6H_{10}O_4(H_2PO_4)_2$. Subsequently it was found that hexosemonophosphates were also formed. It was also reported that whatever sugar was used the hexose was always *fructose* diphosphate in which the fructose was in the active γ or furanose form, and it was therefore supposed until comparatively recently that the significance of this phosphorylation was bringing the hexose into the active form. Recent research has, however, led to a modification of this view.

Phosphorylation does not, of course, involve any breakdown of the hexose molecule. The breaking down of the furanose ring to compounds containing fewer than six carbon atoms must follow phosphorylation and is usually termed glycolysis. There is some variation in the meaning attached to this term as regards the extent of the breakdown covered by it. It is usual to limit glycolysis to the production of compounds containing three carbon atoms from hexose or a phosphorylated product of a hexose, but Turner would prefer the term triosis for this process in plants, as the term glycolysis has been widely used to describe the production of lactic acid from glycogen or hexose in animal tissues and yeast under certain conditions, whereas in plant tissues, including fermenting yeast, the three-carbon atom compounds produced do not normally include lactic acid.

Various 3-carbon atom substances had been suggested from time to time as products of glycolysis, including lactic acid ($CH_3.CHOH.COOH$), methyl glyoxal or pyruvic aldehyde ($CH_3.CO.CHO$), glyceraldehyde ($CH_2OH.CHOH.CHO$) and dihydroxyacetone ($CH_2OH.CO.CH_2OH$), but the discovery by Neuberg and Karczag in 1911 of an enzyme in yeast which effected the removal of carbon dioxide from pyruvic acid with the production of acetaldehyde indicated the probability that the 3-carbon atom pyruvic acid $CH_3.CO.COOH$ was an intermediate in the process of yeast fermentation, that the carbon dioxide produced arose in this way and that acetaldehyde was an intermediate between pyruvic acid and the final product ethyl alcohol. Much subsequent work has supported this view. The enzyme, carboxylase, or more properly decarboxylase because it brings about a decarboxylation, was subsequently found to be of widespread occurrence throughout the plant kingdom. It has already been mentioned that there is considerable evidence for the production of acetaldehyde during fermentation of yeast and the anaerobic respiration of other plants.

A scheme for the production of alcohol and carbon dioxide from hexose which took account of these facts was put forward by Neuberg. According to this hexose was first broken down to methyl glyoxal or pyruvic aldehyde, possibly by way of glyceraldehyde. This was followed by a Cannizzaro reaction in which two molecules of methyl glyoxal were concerned, one being reduced to glycerol, and the other oxidized to pyruvic acid. This was then subjected to the action of carboxylase with production of acetaldehyde and carbon dioxide. The latter was given off and the acetaldehyde, reacting with pyruvic aldehyde produced in glycolysis, gave rise to equimolecular quantities of pyruvic acid and ethyl alcohol. The pyruvic acid was then immediately acted upon by carboxylase. This scheme may be summarized thus:

$$C_6H_{12}O_6 = 2CH_3.CO.CHO + 2H_2O$$
pyruvic aldehyde

$$2CH_3.CO.CHO + 2H_2O = CH_2OH.CHOH.CH_2OH + CH_3.CO.COOH$$
glycerol   pyruvic acid

$$CH_3.CO.COOH = CH_3.CHO + CO_2$$
acetaldehyde

$$CH_3.CO.CHO + CH_3.CHO + H_2O = CH_3.CO.COOH + CH_3.CH_2OH$$
ethyl alcohol

When acetaldehyde has once been formed the second

stage may be eliminated since the pyruvic aldehyde produced in glycolysis at once reacts with acetaldehyde to produce more ethyl alcohol and pyruvic acid which again gives rise to more acetaldehyde. On this scheme only a trace of glycerol will thus be produced, and only ethyl alcohol will accumulate. In support of this scheme is the fact recorded by Neuberg that if the aldehyde is fixed by the addition of a sulphite to the fermenting liquor, so that the last stage is suppressed, there is not only an accumulation of aldehyde sulphite but also of glycerol.

Later work showed that Neuberg's scheme was an over-simplification. The discovery by Embden and his co-workers of phosphoglyceric acid among the products of carbohydrate breakdown and work by Meyerhof and his associates on enzymes present in yeast, indicates that the number of enzymes involved in fermentation is considerably greater than the original Neuberg scheme required, and that phosphorylated compounds are involved until the pyruvic acid stage is reached.

It has already been mentioned that, previous to glycolysis, hexose is combined with phosphate to produce fructose diphosphate. It now appears that the phosphorylation of hexose is not brought about by inorganic phosphate as at one time supposed but by an organic phosphate named adenosine triphosphate. The phosphates of adenosine are substances of the first importance in fermentation and respiration, and it will be appropriate to indicate their nature. Adenosine itself is a nucleoside, that is, a glycoside formed by the union of a pentose sugar with a nitrogen base, the sugar being d-ribose and the base adenine or 6-aminopurine, that is, a substance in which $—NH_2$ replaces a hydrogen atom in the molecule of purine. The formula of adenosine is thus:

$$\begin{array}{c}
N=C-NH_2 \\
| \quad\quad | \\
HC \quad C-N \\
| \quad\quad \|\!\!>\!\!CH \\
N-C-N \!\!\nearrow
\end{array}
\underset{\begin{array}{cccc} | & | & | & | \\ H & H & H & H \end{array}}{\boxed{\begin{array}{c} ———O——— \\ \quad OH \; OH \\ C—C—C—C—CH_2OH \end{array}}}$$

There are three adenosine phosphates. Adenosine monophosphate, or adenylic acid, has the normal structure of an organic phosphate; its formula is thus:

$$\begin{array}{c}
N=C-NH_2 \\
| \quad\quad | \\
HC \quad C-N \\
\| \quad\quad \| \quad\quad CH \\
N-C-N
\end{array}
\begin{array}{c}
\overline{O} \\
\\
OH\ OH \quad\quad\quad O \\
| \quad | \quad\quad\quad \| \\
-C-C-C-C-CH_2-O-P-OH \\
| \quad | \quad | \quad | \quad\quad\quad | \\
H \ H \ H \ H \quad\quad\quad OH
\end{array}$$

It can be briefly represented by the formula

$$A-O-\underset{\underset{OH}{|}}{\overset{\overset{O}{\|}}{P}}-OH$$

The constitutions of adenosine diphosphate and adenosine triphosphate were worked out by Lohmann. Their formulae are now generally written:

$$A-O-\underset{\underset{OH}{|}}{\overset{\overset{O}{\|}}{P}}-O\sim\underset{\underset{OH}{|}}{\overset{\overset{O}{\|}}{P}}-OH \quad \text{and} \quad A-O-\underset{\underset{OH}{|}}{\overset{\overset{O}{\|}}{P}}-O\sim\underset{\underset{OH}{|}}{\overset{\overset{O}{\|}}{P}}-O\sim\underset{\underset{OH}{|}}{\overset{\overset{O}{\|}}{P}}-OH$$

where the symbol $\sim$ indicates an energy-rich phosphate bond. The distinction between energy-poor and energy-rich bonds was made by Lipmann, the former being found in normally constituted organic phosphates where the phosphate grouping is linked to an alcoholic grouping as in hexose phosphates and in adenosine monophosphate while the latter are found where the phosphate is linked to another phosphate grouping or in some other ways which do not concern us here. When compounds of the first group are hydrolysed the energy released amounts to about 2 to 4 Calories per gram molecule, whereas the splitting off of the terminal phosphate residue from an organic phosphate of the second kind is accompanied by a release of from 12 to 15 Calories. For the sake of brevity it is usual to denote adenosine diphosphate and adenosine triphosphate by the symbols ADP and ATP respectively.

We are now in a position to consider the course of fermentation, and so presumably of anaerobic respiration, in detail. It would appear that probably thirteen actions are involved in the degradation of glucose to ethyl alcohol and carbon dioxide.

1. By the action of the enzyme hexokinase, first isolated from yeast by Meyerhof in 1927, the phosphorylation of glucose is effected by the transfer of the terminal phosphate residue of adenosine triphosphate to glucose, with the production of glucose-6-phosphate and adenosine diphosphate:

$$C_6H_{12}O_6 + ATP \rightleftharpoons C_6H_{11}O_5(H_2PO_4) + ADP$$

That phosphorylation is effected by adenosine triphosphate and not inorganic phosphate may be explained on the ground that the phosphorylation requires energy which is provided when the energy-rich bond linking the final phosphate group of the triphosphate is broken, its energy or part of it being utilized in the production of the glucose-6-phosphate.

If starch should form the substrate of fermentation or anaerobic respiration it would appear that glucose-6-phosphate is formed without the intermediate formation of uncombined glucose. From peas and potatoes Hanes obtained a preparation of a phosphorylase enzyme which, in presence of inorganic phosphate, brought about the formation of glucose-1-phosphate, also known as the Cori ester. The glucose-1-phosphate is then, through the action of the enzyme phosphoglucomutase, converted to glucose-6-phosphate. According to Cori and his associates, phosphoglucomutase is of wide distribution in both plants and animals. This enzyme produces an equilibrated mixture of glucose-1-phosphate and glucose-6-phosphate from either.

2. The glucose-6-phosphate is now converted to its isomer fructose-6-phosphate by the action of the enzyme phosphohexose isomerase. This enzyme, first found in muscle extract by Lohmann, acts on either of these

hexosephosphates to produce an equilibrated mixture of the two. There is evidence that the enzyme occurs in yeast and higher plants.

3. The next stage in the process is the production from fructose-6-phosphate of that fructose-1,6-diphosphate first isolated by Harden and Young. This second phosphorylation is also brought about by adenosine triphosphate, the enzyme concerned being phosphohexokinase. The action

$$C_6H_{11}O_5(H_2PO_4) + ATP \rightleftharpoons C_6H_{10}O_4(H_2PO_4)_2 + ADP$$

of this enzyme is thus similar to that of hexokinase, but it has not been purified and it is usual to regard it as a separate enzyme. Fructose-1,6-diphosphate has been isolated by both Tankó and Hanes when an extract of peas is allowed to act on starch in presence of inorganic phosphate.

4. In the next reaction the actual splitting of the hexose grouping is brought about by the enzyme known as aldolase or zymohexase.[1] The products of the reaction are two isomeric 3-carbon atom compounds, dihydroxyacetone phosphate and phosphoglyceric aldehyde (glyceraldehyde-3-phosphate):

$$\begin{array}{c} CH_2.O.H_2PO_3 \\ | \\ COH \\ | \\ CHOH \\ | \\ CHOH \\ | \\ CH \\ | \\ CH_2.O.H_2PO_3 \end{array} \rightleftharpoons \begin{array}{c} CH_2.O.H_2PO_3 \\ | \\ C=O \\ | \\ CH_2OH \\ \\ \text{dihydroxyacetone} \\ \text{phosphate} \end{array} + \begin{array}{c} CHO \\ | \\ CHOH \\ | \\ CH_2.O.H_2PO_3 \\ \\ \text{phosphoglyceric} \\ \text{aldehyde} \end{array}$$

5. The two 3-carbon atom compounds (triosephosphates) produced by the action of aldolase are inter-

[1] The enzyme preparations originally called zymohexase by Meyerhof and Lohmann were later found by them to involve two enzymes, which they then called aldolase and triosephosphate isomerase, aldolase being the enzyme effecting the splitting of the hexose grouping. Warburg and Christian, who obtained the enzyme in a crystalline form, used the term zymohexase as synonymous with aldolase.

convertible to one another by the action of the enzyme triosephosphate (or phosphotriose) isomerase. An equilibrated mixture of the two triosephosphates results from the action of this enzyme, the greater part of the equilibrated mixture consisting of dihydroxyacetone phosphate. The enzyme has been shown to be present in yeast and was prepared in a purified state by Meyerhof and Beck in 1944.

6. It is the phosphoglyceric aldehyde which is utilized in the further degradation processes, but degradation does not occur immediately. The phosphoglyceric aldehyde is oxidized to phosphoglyceric acid, but probably several reactions are involved in the production of the acid from the aldehyde. By the action of the oxidizing enzyme triosephosphate dehydrogenase, which requires the presence of coenzyme 1, diphosphoglyceric acid is produced, but this probably takes place in two stages. It is thought that the first step, which requires the presence of inorganic phosphate, results in the production of some intermediate, but this substance has not been isolated. The following equation might represent the reaction:

$$\begin{array}{c}CH_2.O.H_2PO_3\\|\\CHOH\\|\\CHO\end{array} + H_3PO_4 \rightleftharpoons \begin{array}{c}CH_2.O.H_2PO_3\\|\\CHOH\\|\\CH.O.H_2PO_3\\|\\OH\end{array}$$

7. The intermediate so produced is then oxidized by the transference of two atoms of hydrogen to the coenzyme 1 molecule acting as a hydrogen acceptor, whereby diphosphoglyceric acid is produced and reduced coenzyme 1 formed.

$$\begin{array}{c}CH_2.O.H_2PO_3\\|\\CHOH\\|\\CH.O.H_2PO_3\\|\\OH\end{array} + \overline{Co} \rightleftharpoons \begin{array}{c}CH_2.O.H_2PO_3\\|\\CHOH\\|\\CO.O.H_2PO_3\end{array} + \overline{CoH_2}$$

where $\overline{Co}$ represents a molecule of coenzyme 1.

8. The diphosphoglyceric acid now loses one of its phosphate groupings to adenosine diphosphate, the enzyme catalysing the reaction belonging to a group called by Mrs. Needham and Dixon phosphokinases, which transfer phosphate groups from one molecule to another. This particular enzyme is referred to by Dixon as phosphoglyceric phosphokinase. The reaction may be represented thus:

$$\begin{array}{c} CH_2.O.H_2PO_3 \\ | \\ CHOH \\ | \\ CO.O.H_2PO_3 \end{array} + ADP \rightleftharpoons \begin{array}{c} CH_2.O.H_2PO_3 \\ | \\ CHOH \\ | \\ COOH \end{array} + ATP$$

It may be asked whether there is any explanation forthcoming of why the phosphoglyceric aldehyde should not be oxidized directly to phosphoglyceric acid. The explanation as put forward by Kermack is that such a direct oxidation would involve a considerable loss of energy and would be irreversible. Compounds such as diphosphoglyceric acid, with two phosphate groups, have a relatively high energy content; that is, the second phosphate grouping has an energy-rich bond, so that the energy liberated by dehydrogenation is not lost but, as it were, trapped in the diphosphoglyceric acid. When the monophosphoglyceric acid is ultimately formed the energy is again not lost but passed over to adenosine diphosphate to form adenosine triphosphate which is utilized for supplying energy in phosphorylations and other processes.

9. An isomeric change is now brought about in the phosphoglyceric acid, whereby the phosphate grouping is transferred from the third carbon atom of the acid to the second, that is, 3-phosphoglyceric acid is converted to 2-phosphoglyceric acid:

$$\begin{array}{c} (3)CH_2.O.H_2PO_3 \\ | \\ (2)CHOH \\ | \\ (1)COOH \end{array} \rightleftharpoons \begin{array}{c} (3)CH_2OH \\ | \\ (2)CH.O.H_2PO_3 \\ | \\ (1)COOH \end{array}$$

The enzyme responsible for this action was recognized by Meyerhof and Kiessling in 1935. It is known as phosphoglyceromutase.

10. By withdrawal of water from the 2-phosphoglyceric acid under the action of the enzyme enolase, also recognized by Meyerhof and Kiessling, the enolic form of phosphopyruvic acid (2-phosphoenolpyruvic · acid) is produced:

$$\begin{array}{c} CH_2OH \\ | \\ CH.O.H_2PO_3 \\ | \\ COOH \end{array} \rightleftharpoons \begin{array}{c} CH_2 \\ \| \\ C.O.H_2PO_3 \\ | \\ COOH \end{array} + H_2O$$

11. The phosphate grouping is now removed from the phosphopyruvic acid with the production of pyruvic acid. According to Dixon the action is effected by the agency of another phosphokinase, pyruvic phosphokinase. Adenosine diphosphate acts as a phosphate acceptor with the production of adenosine triphosphate:

$$\begin{array}{c} CH_2 \\ \| \\ C.O.P{-}OH \\ {\diagdown}O \\ | \\ COOH \end{array} \begin{array}{c} OH \\ \\ \\ \\ \end{array} + ADP \rightleftharpoons \begin{array}{c} CH_3 \\ | \\ CO \\ | \\ COOH \end{array} + ATP$$

12. The pyruvic acid is now broken down by the action of the enzyme carboxylase as in Neuberg's earlier scheme, the products being acetaldehyde and carbon dioxide:

$$CH_3.CO.COOH = CH_3.CHO + CO_2$$

As already mentioned, carboxylase was first recognized in yeast by Neuberg and Karczag in 1911, and has since been found in both fungi and higher plants. For its action it requires the presence of a coenzyme termed cocarboxylase which has proved to be the pyrophosphate of thiamin or aneurin (vitamin B_1).

13. With the reaction just described, one of the final products of fermentation or anaerobic respiration, carbon dioxide, is produced and released. The other final product, ethyl alcohol, is presumably formed by the reduction of the acetaldehyde produced in the decarboxylation of pyruvic acid. Two enzymes occur in yeast which will effect this reduction and both require coenzyme 1 for their action. One of these, aldehyde mutase, catalyses a Cannizzaro reaction, that is, a reaction between two molecules of an aldehyde wherein one is reduced and the other oxidized, so that the products when the substrate is acetaldehyde are ethyl alcohol and acetic acid:

$$\begin{array}{c} CH_3.CHO \\ + \\ CH_3.CHO \end{array} + \begin{array}{c} H_2 \\ | \\ O \end{array} \rightleftharpoons \begin{array}{c} CH_3.CH_2OH \\ + \\ CH_3.COOH \end{array}$$

An aldehyde mutase has also been obtained from higher plants but it is said not to require coenzyme 1 for its action.

However, while the action of aldehyde mutase would account for the production of ethyl alcohol, there is no evidence that acetic acid is produced in fermentation so that if it were it would mean that it would have to be transformed immediately to some other substance. If none of this other substance were finally transformed to ethyl alcohol, the ratio of alcohol/carbon dioxide produced would be only 0·5, and where the ratio is less than unity it is a possibility that this enzyme may be operative.

No such complication arises in the consideration of the second enzyme which brings about the reduction of acetaldehyde. This is alcohol dehydrogenase, by the action of which reduced coenzyme 1 provides the hydrogen required for the reduction of the acetaldehyde:

$$CH_3.CHO + \overline{CoH_2} \rightleftharpoons CH_3.CH_2OH + \overline{Co}$$

It will be observed that reduced coenzyme 1 is formed in stage 7 of the scheme of carbohydrate degradation, one molecule being formed for every molecule of diphosphoglyceric acid produced. As this ultimately produces one molecule each of carbon dioxide and ethyl alcohol it will be seen that if the alcohol dehydrogenase action is the final stage in fermentation and anaerobic respiration, not only will the ratio of alcohol/carbon dioxide be unity, but the amounts of coenzyme 1 and reduced coenzyme 1 will remain constant when a steady state of carbohydrate degradation obtains.

Almost every one of the reactions summarized above has been effected outside the plant, and a number of the enzymes involved have been prepared in crystalline form. The working out of the details of the fermentation process must be regarded as one of the major achievements of enzyme chemistry of the last fifteen years.

THE COURSE OF AEROBIC RESPIRATION

In contrast with what is known of the course of fermentation, the evidence of the course of aerobic respiration is fragmentary and, for the most part, indirect. As we have seen, it is generally assumed that the course of the two processes is the same up to a certain stage, possibly the stage at which pyruvic acid is decarboxylated, and that the fate of these products is generally different in presence or absence of an adequate supply of oxygen. There are a number of facts which support this view. In the first place, since anaerobic respiration is not a normal process in most plant organs, it would seem likely that the enzymes concerned in it must fulfil some function in normal life processes. Secondly, phosphate has been shown to be important for respiration; thus Richards found that deficiency of phosphorus led to a reduction in the respiration rate of barley leaves, while James and Arney found that as the phosphate ester content of excised barley embryos fell between the second and fourth days after the beginning of germination the respiration rate also fell. They also found that the respiration of barley embryos provided with an adequate supply of sucrose increased with increase in the phosphate concentration of the culture solution. Thirdly, a number of the intermediate products of fermentation have been obtained from higher plants. Reference has already been made to the work of Tankó and Hanes in which fructose-1,6-diphosphate was shown to be produced from inorganic phosphate by pea flour. Hanes also demonstrated the formation by pea flour of hexose-6-phosphates. Bonner has demonstrated the presence of fructose-1,6-diphosphate, fructose-6-phosphate and glucose-1-phosphate in oat coleoptiles. James, Heard and James also produced evidence to show that phosphoglycerate was among the products when the expressed sap of barley seedlings was incubated with added hexosediphosphate. There was some evidence that triosephosphate might be an intermediate between hexosediphosphate and phosphoglycerate since by adding to the barley sap small quantities of iodoacetate, which inhibits the decomposition of triosephosphate, esters are formed possessing the properties of triosephosphates.

Again James and Norval found that carboxylase was present in barley and that when pyruvic acid was supplied to living barley leaves or embryos the pyruvic acid was broken down and there was an increase in the rate of carbon dioxide output presumably attributable to carboxylase action. Later James and James found that when barley roots were poisoned with reagents which inactivate carboxylase, such as a 0·1 per cent. solution of acetaldehyde or 0·3 per cent. solutions of various aromatic sulphonic acids, pyruvic acid could be recognized in the tissues, while they actually isolated pyruvic acid as the 2,4-dinitrophenyl hydrazone from cut barley leaves treated in the dark with 0·2 per cent. 1-naphthol-2-sulphonic acid. Also Neuberg and Kobel found that phosphoglycerate was converted to pyruvate by preparations of pea and bean, while James, James and Bunting found that addition of sucrose and adenylic acid, or glucose and adenylic acid, to cell-free barley sap brought about the production of readily identifiable pyruvic acid. Hexosediphosphate and phosphoglycerate were also converted into pyruvic acid by the action of barley sap. Pyruvic acid has also been isolated from the onion (*Allium Cepa*) by Bennett and by Morgan, but it has been suggested by Goddard and Meeuse that here it arises from the hydrolysis of alliine to allicine, pyruvic acid and ammonia by the enzyme alliinase and is unconnected with respiration. Bonner found that the addition of pyruvate to oat coleoptile tissue depleted of carbohydrate by removal of the endosperm from the germinated grains led to a 57 to 94 per cent. increase in respiration rate.

There is then some evidence that the course of aerobic respiration is the same as that of fermentation up to the stage in which pyruvic acid is decarboxylated by carboxylase. The widespread presence of this enzyme in higher plants, and James's work to which reference has already been made, suggest that under aerobic, as well as anaerobic conditions, acetaldehyde is produced. At the same time it must be realized that the evidence that the paths of degradation of carbohydrate to the pyruvic acid stage are identical under aerobic and anaerobic conditions is far from complete, and more than one writer has suggested that although similar the paths may not be the same, and that they may even differ in different species.

It has been supposed in the past that from the point where the courses of anaerobic and aerobic respiration differ, under aerobic conditions all the intermediate is converted finally to carbon dioxide and water. There is no reason for assuming that this is generally so. Should the rate of glycolysis be the same under aerobic and anaerobic conditions the ratio of carbon dioxide evolved in aerobic to that evolved in anaerobic respiration should be 3, but only occasionally is this value observed. There is a possibility that the intermediate is only in part oxidized to carbon dioxide and water and that part is built back into the system. As will be shown later, evidence has been advanced to show that this is what actually happens.

However this may be, it is clear that the final stages in aerobic respiration must involve oxidations, and it can be regarded as certain that enzyme systems are concerned. It will therefore be appropriate to consider briefly the various oxidizing enzymes met with in plants, although it is still very problematic which of these are actually concerned in respiration.

There is no generally accepted classification of oxidizing enzymes, but for our purpose they may be regarded as falling into three groups, dehydrogenases, peroxidases and oxidases. Two other enzymes, catalase and carbonic anhydrase, which are not, strictly speaking, oxidizing enzymes, also deserve mention.

1. *Dehydrogenases*

These enzymes act by the transference of hydrogen from one substance (the hydrogen donor or donator) to another (the hydrogen acceptor) whereby the former is oxidized and the latter reduced. Reference has already been made to the action of two of these in dealing with the anaerobic breakdown of carbohydrate. They have been divided into three groups: (*a*) those requiring coenzyme 1 or the nearly related coenzyme 2[1] for their action, (*b*) those possessing an active (prosthetic) flavin group and which do not require a coenzyme, and (*c*) those which reduce cytochrome *c* and which also do not require a coenzyme. In these enzymes the hydrogen acceptor is the coenzyme, flavin or cytochrome as the case may be, while the reduced coenzyme, flavin or cytochrome will act as a hydrogen donor (cf. the actions of triosephosphate dehydrogenase and alcohol dehydrogenase mentioned on pp. 105 and 108). Some authorities include among dehydrogenases those enzymes which

[1] Coenzyme 2 is similar to coenzyme 1, but contains three phosphate groupings instead of two. It is thus triphosphopyridine nucleotide (TPN).

transfer hydrogen to molecular oxygen. Where, however, as with a number of plant oxidizing enzymes, the oxidation can be effected *only* by molecular oxygen, it is convenient to consider these as a separate class, the oxidases.

2. *Peroxidases*

These enzymes bring about the oxidation of a number of phenolic compounds, including catechol, pyrogallol and cresols, in presence of hydrogen peroxide. They are probably universally distributed throughout the plant kingdom for they have been found in nearly all higher plants where they have been sought. Theorell obtained two peroxidases from horse-radish, one of which he was able to prepare in crystalline form. This peroxide he was able to split into a protein component and an active component which proved to be the iron-containing hematin which could therefore be regarded as coenzyme or active prosthetic group according to whether the hematin is to be regarded as a separate substance or part of the enzyme molecule.

3. *Oxidases*

These are the enzymes which only oxidize by means of molecular oxygen. The best known of these are catechol (or polyphenol) oxidase, cytochrome (or indophenol) oxidase and ascorbic acid oxidase.

(i) *Catechol oxidase*.—The term oxidase was first applied by Chodat and Bach to the enzyme system present in many plants which produces a blue colour in a tincture of guaiacum gum, the blue colour being an oxidation product of guaiaconic acid, a phenolic constituent of the gum. In 1911 Miss Wheldale showed that all plants giving the guaiacum reaction contained a substance or substances possessing the catechol grouping

and later she showed that the oxidation of the catechol

compounds was effected by an enzyme, and that the products of this oxidation would effect the blueing of a tincture of guaiacum. To this enzyme she gave the name oxygenase, but as it catalyses the oxidation of substances of the catechol type, it is appropriately called catechol oxidase or polyphenol oxidase. The mode of action is not quite clear, but it would appear that both hydrogen peroxide and a quinone are produced:

$$\text{C}_6\text{H}_4(\text{OH})_2 + \text{O}_2 = \text{C}_6\text{H}_4\text{O}_2 + \text{H}_2\text{O}_2$$

Not only is the hydrogen peroxide then capable of bringing about further oxidations in presence of peroxidase, but the quinone can also bring about secondary oxidations such as the oxidation of a monohydric phenol to a dihydric phenol:

p-cresol → homocatechol

The diphenol so produced can then be oxidized to a quinone by the catechol oxidase which can thus effect indirectly the oxidation of monophenols to quinones.

Catechol oxidase belongs to a group of enzymes effecting the oxidation of phenolic compounds, which are known on that account as phenolases. Another is laccase, which oxidizes the diphenols uroshiol and laccol present in the latex of the lac trees *Rhus succedanea* and *Rhus nucifera* respectively.

Tyrosinase, which has been known for more than half a century, is a phenolase which catalyses the oxidation of the phenolic amino-acid tryosine and some other phenolic compounds. Because these include monohydric phenols it has also been known as monophenol oxidase. It appears to be distinct from catechol oxidase, but some authorities use the terms monophenol oxidase, polyphenol oxidase, catecholase and tyrosinase as synonymous with phenolase, but it would seem likely that we have here to do with a group of similar enzymes rather than a single enzyme.

It is important to note that catechol oxidase, and probably all phenolases, are copper enzymes; that is, they are proteins with a small amount of copper in their molecules. The amount of copper has been found by various workers to be from 0·2 to 0·3 per cent. of the whole enzyme molecule.

It is also important to note that the action of these enzymes can be inhibited by a number of reagents which include cyanides, sulphides, azides and carbon monoxide. The inhibition of the enzyme by carbon monoxide is unaffected by light.

(ii) *Cytochrome oxidase*.—In 1925 and subsequent years Keilin showed the presence in yeast and animal tissues of a number of hematin compounds which have definite oxidative properties. In yeast Keilin distinguished four such compounds, an unbound protohematin and three hematin compounds which he called a', b' and c' cytochrome respectively. The first two are unstable, but cytochrome c' is very stable. It is a hematin-protein compound and exists in a reduced and oxidized form, but is not itself autoxidizable.

The enzyme effecting the oxidation of reduced cytochrome c is cytochrome (or indophenol) oxidase. More than fifty years ago an enzyme of both plants and animals was recognized which brought about the production of indophenol blue from a mixture of α-naphthol and dimethyl-*p*-phenylenediamine. This enzyme was called indophenol oxidase, and it was found to bring about the oxidation of a number of phenolic substances including phenylenediamine, hydroquinone and catechol. It was shown by Keilin to be distinct from catechol oxidase which can only oxidize phenylenediamine in presence of a catechol compound. Also it does not blue a tincture of guaiacum. Later research has produced strong evidence that indophenol oxidase only brings about the oxidation of reduced cytochrome c and that the phenolic compounds are then oxidized by the oxidized cytochrome c which is thereby reduced. For this reason the enzyme is now generally known as cytochrome oxidase.

Cytochrome oxidase is probably a hemin-protein and so contains iron, the iron content of the purified enzyme being about 0·3 per cent. Like catechol oxidase, cytochrome oxidase is inhibited by cyanides, sulphides and azides. It is inhibited by carbon monoxide in the dark, but the inhibition is removed by exposure to light. It retains its activity in very low oxygen concentrations.

(iii) *Ascorbic acid oxidase.*—Ascorbic acid, hexuronic acid or vitamin C is widely distributed throughout the plant kingdom. An enzyme effecting the oxidation of this substance was found in 1931 by Szent-Györgyi in leaves of cabbage and has since been found in the tissues of other plants, including species of Leguminosae, Umbelliferae and Cucurbitaceae. By the action of the enzyme *l*-ascorbic acid is oxidized to dehydroxyascorbic acid:

$$\begin{array}{c}CO\\|\\C-OH\\\|\\C-OH\\|\\H-C\\|\\HO-C-H\\|\\HO-C=H_2\end{array}\Bigg] O \;+\; O \;=\; \begin{array}{c}CO\\|\\C=O\\|\\C=O\\|\\H-C\\|\\HO-C-H\\|\\HO-C=H_2\end{array}\Bigg] O \;+\; H_2O$$

A number of substances related to *l*-ascorbic acid are also oxidized by ascorbic acid oxidase. These include *l*-glucoascorbic acid, *l*-galactoascorbic acid, reductic acid and reductone, the last two compounds having the formulae

```
    CO─────┐
    │      │
    C─OH   │
    ‖      CH₂
    C─OH   │
    │      │
    H₂C────┘
```
reductic acid

```
    H─CO
    │
    C─OH
    ‖
    H─C─OH
```
reductone

There is some evidence that ascorbic acid oxidase, like catechol oxidase, is a copper protein compound. The action of the enzyme is inhibited by cyanide, but according to James is not affected by azide or carbon monoxide.

(iv) *Glucose oxidase.*—An enzyme bringing about the oxidation of glucose to gluconic acid by oxygen was found by Müller in 1925 in *Aspergillus niger*. The same or a similar enzyme was recorded twenty years later by Coulthard and his associates in *Penicillium notatum* and *P. resticulosum*. This enzyme was named glucose oxidase by Müller and notatin by Coulthard and his co-workers.

Glucose oxidase is not inhibited by cyanides, sulphides or azides.

4. *Catalase*

This enzyme is widely distributed throughout the plant kingdom and quite possibly is present in all plants. It brings about the decomposition of hydrogen peroxide into water and oxygen. As the oxygen is given off in the molecular state, catalase is not generally regarded as an oxidizing enzyme and its function is supposed to be the destruction of hydrogen peroxide which may be produced as a by-product of metabolism. However, Keilin and Hartree concluded that catalase could take part in oxidations by catalysing the oxidation of a number of alcohols by the hydrogen peroxide produced in some oxidase actions. Thus xanthine oxidase, which occurs in animal tissues, catalyses the oxidation of hypoxanthine by oxygen to uric acid, the oxygen being reduced to hydrogen peroxide. If catalase and ethyl alcohol are added to this system the alcohol is oxidized to actealdehyde :

$$CH_3.CH_2OH + H_2O_2 = CH_3.CHO + 2H_2O$$

Like cytochrome oxidase and peroxidase, catalase contains iron. It is inactivated by cyanides, sulphides, azides and hydroxylamine. It has been prepared in crystalline form.

5. *Carbonic Anhydrase*

This enzyme catalyses the decomposition of carbonic acid into carbon dioxide and water and so might function in the release of carbon dioxide from plant cells. So far it has not been recorded for many plant tissues, but has been found in the chloroplasts and other parts of the leaf cells of *Trifolium pratense* and *Onoclea sensibilis*. The enzyme is a zinc-protein compound and is inhibited in its action by cyanides, sulphides, azides, salts of heavy metals and carbon monoxide.

With this brief review of oxidizing enzymes in plants we are in a better position to consider the oxidation phase in aerobic respiration. It has already been shown that there is some evidence to justify the Pfeffer-Kostychev theory of a common course of carbohydrate degradation under aerobic and anaerobic conditions up to a certain point. With regard to the actual position of this point there is no generally accepted opinion. The wide distribution of carboxylase in plants suggests that the starting-point for oxidation might be acetaldehyde. James has suggested the possibility of oxidation interrupting the course of glycolysis at more than one stage, but since aerobic respiration is adversely affected by iodoacetate (cf. p. 89), which inhibits the action of triosephosphate dehydrogenase, he thought that degradation of carbohydrate proceeded in air as under anaerobic conditions at least as far as the substrate of this enzyme.

The starting-point for the oxidation phase in aerobic respiration might then be a phosphoglyceric acid, phosphopyruvic acid, pyruvic acid or acetaldehyde. For animal tissues pyruvic acid has been regarded with favour as the starting-point of the oxidation process, and more than one scheme has been proposed of the mechanism of this oxidation. The best known of these is perhaps the Krebs cycle, according to which a molecule of pyruvic acid reacts with one of oxalacetic acid to produce a molecule of citric acid and one of carbon dioxide. By the action of the enzyme aconitase the citric acid is converted through cis-aconitic acid to isocitric acid, which then by the action of isocitric dehydrogenase is oxidized to α-keto-β-carboxyglutaric acid which decomposes spontaneously to α-ketoglutaric acid and carbon dioxide. The α-ketoglutaric acid is then oxidized by the action of α-ketoglutaric dehydrogenase to succinic acid and carbon dioxide. By the action of succinic dehydrogenase and fumaric dehydrogenase, fumaric acid and malic acid are produced by successive oxidations. By the action of malic acid dehydrogenase oxalacetic acid is then reformed from malic acid. This series of reactions can be represented thus:

(1) by oxidation and decarboxylation:

$$\begin{array}{c} CH_3 \\ | \\ CO \\ | \\ COOH \end{array} + \begin{array}{c} COOH \\ | \\ CH_2 \\ | \\ CO \\ | \\ COOH \end{array} + X + H_2O \rightleftharpoons \begin{array}{c} COOH \\ | \\ CH_2 \\ | \\ HO-C-COOH \\ | \\ CH_2 \\ | \\ COOH \end{array} + CO_2 + XH_2$$

pyruvic acid   oxalacetic acid   citric acid

The production of citric acid probably involves more than one reaction. The mechanism is doubtful and the hydrogen acceptor X is unknown but possibly coenzyme A which contains the grouping of pantothenic acid, is involved.

(2) by aconitase:

$$\begin{array}{c}\text{COOH}\\|\\\text{CH}_2\\|\\\text{HO}-\text{C}-\text{COOH}\\|\\\text{CH}_2\\|\\\text{COOH}\end{array} \rightleftharpoons \begin{array}{c}\text{COOH}\\|\\\text{CH}\\\|\\\text{C}-\text{COOH}\\|\\\text{CH}_2\\|\\\text{COOH}\end{array} + H_2O \rightleftharpoons \begin{array}{c}\text{COOH}\\|\\\text{CHOH}\\|\\\text{H}-\text{C}-\text{COOH}\\|\\\text{CH}_2\\|\\\text{COOH}\end{array}$$

citric acid · · · · · · · cis-aconitic acid · · · · · · · l-isocitric acid

(3) by isocitric dehydrogenase:

$$\begin{array}{c}\text{COOH}\\|\\\text{CHOH}\\|\\\text{H}-\text{C}-\text{COOH}\\|\\\text{CH}_2\\|\\\text{COOH}\end{array} + \overline{Co} \rightleftharpoons \begin{array}{c}\text{COOH}\\|\\\text{CO}\\|\\\text{H}-\text{C}-\text{COOH}\\|\\\text{CH}_2\\|\\\text{COOH}\end{array} + \overline{CoH_2}$$

l-isocitric acid · · · · · · · a-keto-β-carboxyglutaric acid

In animal tissues this dehydrogenase requires coenzyme 2 as hydrogen acceptor, but according to Andersson, in plants it requires coenzyme 1.

(4) by decarboxylation, possibly by a carboxylase:

$$\begin{array}{c}\text{COOH}\\|\\\text{CO}\\|\\\text{H}-\text{C}-\text{COOH}\\|\\\text{CH}_2\\|\\\text{COOH}\end{array} \rightleftharpoons \begin{array}{c}\text{COOH}\\|\\\text{CO}\\|\\\text{CH}_2\\|\\\text{CH}_2\\|\\\text{COOH}\end{array} + CO_2$$

α-keto-β-carboxyglutaric acid · · · α-ketoglutaric acid

(5) by α-ketoglutaric dehydrogenase:

$$\begin{array}{c}\text{COOH}\\|\\\text{CO}\\|\\\text{CH}_2\\|\\\text{CH}_2\\|\\\text{COOH}\end{array} + H_2O + Y \rightleftharpoons \begin{array}{c}\text{COOH}\\|\\\text{CH}_2\\|\\\text{CH}_2\\|\\\text{COOH}\end{array} + CO_2 + YH_2$$

α-ketoglutaric acid · · · · · · · succinic acid

The hydrogen acceptor for this action is unknown, but it is supposed also to require the presence of cocarboxylase.

(6) by succinic dehydrogenase:

$$\begin{array}{c}\text{COOH}\\|\\\text{CH}_2\\|\\\text{CH}_2\\|\\\text{COOH}\end{array} + \tfrac{1}{2}\text{O}_2 \rightleftharpoons \begin{array}{c}\text{COOH}\\|\\\text{CH}\\||\\\text{CH}\\|\\\text{COOH}\end{array} + \text{H}_2\text{O}$$

succinic acid   fumaric acid

Under anaerobic conditions methylene blue, thionine and several other dyes can act as hydrogen acceptors in the action of succinic dehydrogenase, but under aerobic conditions the hydrogen is accepted by oxygen through the action of the cytochrome oxidase system. The action is not really understood, but it could be explained by the hydrogen first reducing the cytochrome, the reduced cytochrome then being converted back to cytochrome c by the action of cytochrome oxidase in presence of oxygen which accepts the hydrogen from the reduced cytochrome. The net result is thus expressed by equation 6 above.

(7) by fumarase:

$$\begin{array}{c}\text{COOH}\\|\\\text{CH}\\||\\\text{CH}\\|\\\text{COOH}\end{array} + \text{H}_2\text{O} \rightleftharpoons \begin{array}{c}\text{COOH}\\|\\\text{CHOH}\\|\\\text{CH}_2\\|\\\text{COOH}\end{array}$$

fumaric acid l-malic acid

(8) by malic dehydrogenase:

$$\begin{array}{c}\text{COOH}\\|\\\text{CHOH}\\|\\\text{CH}_2\\|\\\text{COOH}\end{array} + \overline{\text{Co}} \rightleftharpoons \begin{array}{c}\text{COOH}\\|\\\text{CO}\\|\\\text{CH}_2\\|\\\text{COOH}\end{array} + \overline{\text{CoH}}_2$$

l-malic acid   oxalacetic acid

With the re-formation of oxalacetic acid the cycle is complete and this acid can again condense with more pyruvic acid. The net result of this cycle of operations is the loss of pyruvic acid, three molecules of carbon dioxide being produced for every molecule of pyruvic acid which disappears. There are also formed four molecules of reduced coenzyme, but since reduced coenzyme does not accumulate there must be a mechanism for their oxidation. This presumably is effected through the action of an oxidase or oxidases, so that the hydrogen accepted by the coenzyme is ultimately passed on to oxygen. Thus reduced coenzyme 1 is oxidized by an enzyme diaphorase in which the active group is a flavin which accepts hydrogen from the reduced coenzyme 1. It is generally assumed that the hydrogen is then passed on to oxygen through the action of the cytochrome oxidase system. The oxidation of reduced coenzyme 1 can then be summarized thus:

$$\overline{Co}H_2 + D = \overline{Co} + DH_2$$
$$DH_2 + Cy = D + CyH_2$$
$$CyH_2 + \tfrac{1}{2}O_2 = Cy + H_2O$$

There is also a molecule of reduced coenzyme 1 produced in stage 7 of the breakdown of glucose (p. 105). Under anaerobic conditions this reduced coenzyme is converted back to the oxidized form in the final stage (13) of fermentation or anaerobic respiration. Under aerobic conditions, however, this final stage is suppressed. Hence altogether in aerobic respiration there will be five molecules of reduced coenzyme oxidized by oxygen through the action of oxidases for every molecule of pyruvic acid formed and destroyed. With the half-molecule of oxygen used in the conversion of succinic acid to fumaric acid this means that six half-molecules of oxygen or $3O_2$ will be absorbed for every molecule of pyruvic acid formed in the respiration process and for the three molecules of carbon dioxide produced. As a molecule of water is produced for each of these molecules of coenzyme oxidized, an examination of the equations given on the preceding pages will show that altogether three molecules of water are produced for each molecule of pyruvic acid lost. Since each molecule of glucose gives rise to two molecules of pyruvic acid, the net change over the whole respiration process is thus

$$C_6H_{12}O_6 + 6O_2 = 6CO_2 + 6H_2O$$

It will be observed that the oxidation, and the energy release accompanying it, does not occur in a single reaction, but, like the carbon dioxide evolution, in a number of steps. The oxidase catalysing the final step in the transfer of hydrogen from metabolites to oxygen is now often referred to as a terminal oxidase.

It is natural that evidence should have been sought of a similar system in the respiration of plant tissues. That aerobic respiration in plants might involve such a cycle of changes was suggested by Bonner and Wildman in 1945. Certainly the presence of a number of the supposed intermediates of the Krebs cycle and of the enzyme systems involved have been demonstrated in plants. The occurrence of pyruvic acid has already been mentioned. Apart from the fruits of many species in which citric acid is abundant, this acid has been recognized in other plant organs, as for instance potato tubers by Guthrie and cotton by McCall and Guthrie and by Ergle and Eaton. Isocitric acid, as well as occurring in some fruits including the blackberry, has been isolated by Pucher, Abrahams and Vickery from leaves of *Bryophyllum*. Both succinic acid and malic acid have been shown to be widely distributed throughout the plant kingdom. Oxalacetic acid and α-ketoglutaric acid are regarded as of great importance in nitrogen metabolism, and in relation to this have been recorded as present in pea plants (*Pisum*) by Virtanen and Laine and by Damodaran and Nair, but so far their widespread occurrence in plants has not been demonstrated.

Among the enzyme systems involved in the Krebs cycle aconitase, isocitric dehydrogenase, succinic dehydrogenase, fumarase and malic dehydrogenase have all been recognized in plants; so have the oxidases described earlier (pp. 113–17). There is thus at least a possibility that the oxidation of pyruvic acid in plant respiration may follow a similar course to that hypothesized in the Krebs cycle. It should be realized, however, that the evidence in favour of such a view is very indirect. Indeed, James's view that there might be more than one starting-point for the oxidation phase of respiration has already been mentioned, and it must be decided that the actual course of the oxidative breakdown of the products of glycolysis in plants is still problematic.

The recent investigations of Vennesland and his co-workers on the action of a carboxylase effecting the decarboxylation of oxalacetic acid may prove of interest in relation to the oxidation stages in plant respiration. They found that oxalacetic carboxylase is probably widely distributed in plants. The action of this enzyme is to produce pyruvic acid from oxalacetic acid:

$$\begin{array}{c} COOH \\ | \\ CH_2 \\ | \\ CO \\ | \\ COOH \end{array} \rightleftharpoons \begin{array}{c} CH_3 \\ | \\ CO \\ | \\ COOH \end{array} + CO_2$$

The enzyme appears always to be associated with malic dehydrogenase which effects the oxidation of malic acid to oxalacetic acid (cf. p. 121). The action of the two enzymes working together could be to produce malic acid from pyruvic acid, or vice versa. There is thus the possibility of the interconversion in plant tissues of three of the reactants in the Krebs cycle, but whether these actions are actually concerned in plant respiration cannot be judged at present.

It can, however, be reasonably supposed that the final transfer of hydrogen to oxygen is effected by oxidase action. In recent years a number of investigators have attempted to discover which of the oxidases are concerned in aerobic respiration in plants, but the results obtained have led to conflicting conclusions. Catechol oxidase, cytochrome oxidase and ascorbic acid oxidase have all been held as controlling respiration.

Evidence of the participation of the various plant oxidases in respiration has mainly been sought in three ways: (1) by determining the presence or absence of the particular oxidases, (2) by examining the effects of known inhibitors of the oxidases on respiration, and (3) by observing the effect of adding a substrate of the oxidase to tissue. It seems to us that these kinds of evidence are not above suspicion. The presence of an oxidase is not sufficient evidence of its action as a terminal oxidase in respiration, while although an enzyme has not been found in a tissue it may be there all the same. Also, in such a complex mechanism as the living cell it is possible that such poisons as cyanides may affect other cell constituents besides oxidases and there may result effects on cell processes, including respiration, not directly due to the influence of the poison on the enzyme action it is known to inhibit *in vitro*. Similarly an oxidase substrate, such as catechol, if present in excess, might act as a protoplasmic poison.

From the results of work with potato tuber tissue Boswell and Whiting concluded that catechol oxidase was concerned with the respiration of potatoes. They found that addition of a small quantity of a 0·04 M solution of catechol to thin slices of potato tuber resulted in a considerable increase in respiration intensity, the rate of both oxygen uptake and carbon dioxide evolution rising, the increase being followed by a fall to a value much below the initial rate. The fall was attributed to inhibition of catechol oxidase action by an oxidation product of catechol. With tissue slices two cells thick, in which presumably the catechol will diffuse into all the cells, the final respiration rate was found to be one-third of the initial rate before the addition of catechol. This result led Boswell and Whiting to conclude that two-thirds of the total normal respiration of potato tuber tissue was controlled by catechol oxidase and the remaining third by some other system. Baker and Nelson, however, considered that the fall in respiration rate observed by Boswell and Whiting was not due to inhibition of catechol oxidase by a catechol product, since they observed an even greater fall on addition of 4-tertiary butyl alcohol, which scarcely inhibits the action of catechol oxidase at all. These investigators reported, however, that a reduction in respiration rate, as measured by oxygen uptake, of 85 per cent., occurred when slices of potato were subjected to the action of inhibitors of catechol oxidase, such as potassium cyanide and 4-nitrocatechol, and 85 per cent. they regarded as near enough to 100 per cent. to make it seem probable that the whole of the respiration of potato tuber tissue was controlled by catechol oxidase. Later Boswell found that the actual amount of residual respiration after treatment of the tissue with catechol depended on the condition of the tissue, and he thought the difference between the found values of 67 per cent. and 85 per cent. reduction in respiratory activity was related to tissue differences.

Schade and his collaborators considered that catechol was a protoplasmic poison which completely disrupted the normal metabolism of the cell and they therefore disputed the conclusion that the fall in respiration of potato slices as a result of adding catechol was due to an inhibition of catechol oxidase by an oxidation product of catechol. Their own experiments led them to conclude that catechol oxidase was not a terminal oxidase in potato respiration but that two other oxidases were involved. Since the respiration of potato slices was partially inhibited by carbon monoxide and the inhibition removed by light it was concluded that an enzyme concerned was cytochrome oxidase. Support for this was obtained by the use of homogenates of the potato tissue. Addition of cytochrome c to these brought about an increased oxidation of p-phenylenediamine (cf. p. 115) and of ascorbate. The increment of oxidation was inhibited by carbon monoxide, the inhibition being removed by light. It was not reduced in atmospheres containing a low partial pressure of oxygen.

Since, however, the respiration of potato tissue slices is partially inhibited by low oxygen concentration, Levy and Schade concluded that a second oxidase also functioned in potato respiration. They found that homogenates without added cytochrome c oxidized p-phenylenediamine and ascorbate, the oxidation requiring a relatively high partial pressure of oxygen.

Boswell, while accepting the evidence for the presence of cytochrome oxidase in the potato, vigorously combated the conclusion of Schade and his co-workers that catechol oxidase is not a terminal oxidase in potato respiration, and from a consideration of their results Boswell concluded that their second enzyme is indeed catechol oxidase.

Miss Hackney considered that in apples also catechol oxidase was concerned with respiration. Various phenolic substances such as catechol, p-cresol and protocatechuic acid when added to the medium surrounding thin slices of apple brought about an increase in the rate of oxygen uptake and sometimes in the rate of carbon dioxide evolution. Resorcinol and potassium cyanide, inhibitors of catechol oxidase, both inhibited the respiration of apple slices. It should be noted, however, that cyanide also inactivates cytochrome oxidase.

In animal tissues the favoured view is that cytochrome oxidase is the terminal oxidase in respiration. Evidence that this oxidase might function in a similar way in the roots and leaves of the carrot was produced by Marsh and Goddard. They found that the respiration of carrot root slices was reduced by treatment of these with potassium cyanide, a maximum fall to about 20 per cent. of the original rate being produced by cyanide in a concentration of 3×10^{-4} M; further increase in cyanide concentration brought no further decrease in respiration rate. A similar inhibition was produced by sodium azide and by carbon monoxide, the inhibition of the last being reversible by light. This last observation suggests that cytochrome oxidase and not catechol oxidase (cf. p. 115) is responsible for 80 per cent. of the respiration of carrot root.

In work on similar lines with excised wheat embryos Brown and Goddard concluded that a large proportion of the respiration of the wheat embryo depended on the action of cytochrome oxidase. A similar conclusion has been drawn with regard to wheat and rice seedlings by Taylor, barley seedlings by Merry and Goddard and by Machlis, and oat coleoptiles by Bonner.

Reference has already been made to the conclusion of Schade and his co-workers that cytochrome oxidase is a terminal oxidase in potato tuber tissue. Stenlib concludes, from the effect on the respiration of carrot leaves in presence of sodium azide, 2,4-dinitrophenol and o-phenanthroline, that an enzyme containing a heavy metal is a terminal oxidase in these leaves, but apart from rather querying the function of catechol oxidase in this way, refrains from making a more precise suggestion regarding the identity of the oxidase.

It has also been suggested that ascorbic acid oxidase may play a part in respiration. James and his co-workers found that cyanide inhibited respiration in barley shoots, and that the addition of ascorbic acid to the juice of barley seedlings led to a considerable increase in the absorption of oxygen and that this was completely inhibited by cyanide. The sap contained neither catechol nor cytochrome, the oxidation did not appear to be due to peroxidase, and James concluded that the enzyme responsible was ascorbic acid oxidase. The ascorbic acid oxidase effects, of course, the transfer of hydrogen from reduced ascorbic acid to oxygen, so that the oxidized form is produced; the reduction of the oxidized form James thought was then effected by a dehydrase action in which an α-hydroxy acid was the hydrogen donator, this being reduced to an α-ketonic acid:

$$A + R.CHOH.COOH \rightleftharpoons AH_2 + R.CO.COOH$$
$$AH_2 + \tfrac{1}{2}O_2 \rightleftharpoons A + H_2O$$

The ascorbic acid oxidase would thus be playing much the same role as cytochrome oxidase is supposed to play in bringing about the oxidation of reduced coenzyme in the Krebs cycle.

Miss Hackney thought that ascorbic acid oxidase, as well as catechol oxidase, might be a terminal oxidase in the respiration of apple slices. She found that both oxygen absorbed and carbon dioxide given out were increased by addition of ascorbic acid in low concentrations, the oxygen absorbed being much more than could be accounted for by the ascorbic acid added, thus suggesting that the ascorbic acid formed part of an oxidase system concerned in respiration. As the same effect was obtained by adding the phenolic compound protocatechuic acid, a substrate of catechol oxidase, and as the effects were additive, she concluded that both catechol oxidase and ascorbic acid functioned as terminal oxidases in the respiration of the apple.

Boswell's results with swedes, in which catechol oxidase is absent, indicate that in these also ascorbic acid oxidase may be one of the terminal oxidases in respiration.

So far the enzymes considered to act as terminal oxidases, catechol oxidase, cytochrome oxidase and ascorbic acid oxidase are all metallo-enzymes containing either copper or iron and their action is inhibited by cyanide. James and Beevers have found, however, that the very rapid respiration of the spadix of *Arum* species is not significantly inhibited by 0·001 M cyanide, neither is it affected by carbon monoxide which inhibits the action of catechol oxidase and cytochrome oxidase, nor by diethyldithiocarbamate which inhibits the action of catechol oxidase and ascorbic acid oxidase. Further, none of these enzymes could be found in the spadix. Some indication was obtained that the oxidation stage in the respiration of the *Arum* spadix depended on a flavoprotein enzyme.

The evidence then, as far as it goes, suggests that aerobic respiration follows the same course as fermentation as far as the production of pyruvic acid, and that the oxidation of this may be brought about by a series of reactions similar to those of the Krebs cycle hypothesized for animal tissues. The widespread occurrence of carboxylase in plants does, however, suggest that acetaldehyde may be a starting-point for the oxidation stages in respiration, while other intermediates in fermentation are not ruled out as possible starting-points for this. As regards the oxidases concerned in the final transfer of hydrogen to oxygen there is evidence of a kind that catechol oxidase, cytochrome oxidase and ascorbic acid oxidase may all function in this way.

THE PASTEUR EFFECT AND OXIDATIVE ANABOLISM

Attention has already been drawn to what is now generally known as the Pasteur effect, namely, the effect of oxygen in reducing the apparent rate of glycolysis. The Pasteur effect is generally assumed to operate when the ratio of carbon dioxide evolved under anaerobic conditions is more than one-third of that evolved under aerobic conditions. The reason for this is immediately apparent from an inspection of the two equations:

$$C_6H_{12}O_6 + 6O_2 = 6CO_2 + 6H_2O$$
$$C_6H_{12}O_6 = 2CO_2 + 2C_2H_5OH$$

The assumption is only justified where the products of anaerobic respiration are wholly carbon dioxide and alcohol. Where, as in potato tubers, little or no alcohol is produced, and the actual products are unknown, a Pasteur effect could still be inferred if the ratio of anaerobic respiration to aerobic respiration exceeded unity. By actual determinations of the loss of carbohydrate from apples and oranges in air and in nitrogen

Fidler showed that the rate of carbohydrate loss is indeed less in air than in nitrogen.

The Pasteur effect has been observed in a wide range of plant material including roots of beet, mangold and carrot, artichoke tubers and a number of fruits, including tomatoes, mangoes, guavas and apples.

There appear to be at least three possibilities. Firstly, the effect of oxygen might be actually to reduce the rate at which the carbohydrate substrate is consumed, secondly, the rate of glycolysis might not itself be reduced but there might be other carbon-containing products of oxidation besides carbon dioxide. Both these views have been assumed by writers on the subject, but generally without adequate evidence, which is, indeed, difficult to obtain. It has been suggested that such evidence would be provided if the rate of substrate consumption were determined as well as the quantities of carbon dioxide and alcohol produced, but even this would not settle the question if one of the processes involved in aerobic respiration were a re-synthesis of sugar. However, work in which the loss of substrate under anaerobic conditions, and the production of alcohol and carbon dioxide were actually measured, suggested a third possibility, namely, that under such conditions additional carbon dioxide might be produced from other sources such as proteins or organic acids (cf. p. 65).

On the assumption that oxygen reduces the actual rate of glycolysis, it has been suggested that oxygen inhibits one or other of the various reactions in the anaerobic breakdown of sugar. There is, however, little direct evidence available of such inhibition.

The second interpretation of the Pasteur effect was rendered familiar to botanists by F. F. Blackman's analysis of data of the respiration of apples under aerobic and anaerobic conditions published in 1928. Blackman had then made the assumption, which at the time appeared to be generally accepted, that the action of the zymase complex of enzymes was independent of oxygen.

Where, then, as in apples, the rate of carbon dioxide output on switching over from an atmosphere of air to one of nitrogen did not fall to as low as one-third of the previous rate, it followed that in air only part of the intermediate substances formed as a result of glycolysis

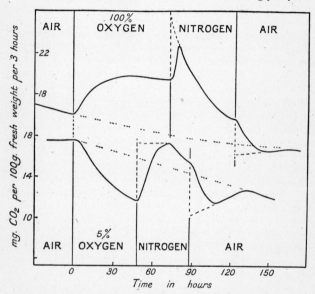

FIG. 10.—Curves showing the effect on carbon dioxide output of Bramley's Seedling apples, produced by atmospheres containing various concentrations of oxygen
(*After* Blackman)

were oxidized completely to carbon dioxide and water, while the rest had some other fate. No new substance appeared to accumulate in the tissues, and the conclusion was therefore drawn that in presence of oxygen part of the products of glycolysis were worked back into the system, whether to the hexose or to some intermediate stage could not be said. To this process the name 'oxidative anabolism' was given. Proof of such a synthesis of carbohydrate associated with oxidation had previously been given by Meyerhof for muscle.

In considering Blackman's conclusions and the reasons which led him to them, let us first note the effect produced on respiration of apples by changing the oxygen content of the environment. When air is replaced by pure oxygen the respiration rate increases, the rate in oxygen being 1·4 times the rate in air, while when normal air is replaced by air containing only 5 per cent. oxygen the respiration rate falls to 0·7 that in normal air, or less. The transition to the new rate is, however, slow in both cases, lasting at least 45 hours (cf. Fig. 10). Now it seems most unlikely that it would take so long a time for a pure oxidation rate to adjust itself after an increase in oxygen concentration, and it is therefore concluded that oxygen concentration must influence the rate of some earlier stage in the process. As it is assumed that the action of the zymase complex of enzymes is independent of oxygen it is presumed that oxygen concentration must affect a pre-glycolytic stage, with the result that increase in oxygen concentration brings about an increase in the concentration of the effective substrate for glycolysis. As this increase in the concentration of the substrate is maintained it means that increase in oxygen concentration results in an increase in the rate of production of the substrate from pyranose sugars or reserve carbohydrates.

That oxygen concentration influences the rate of production of the effective substrate for glycolysis from normal hexoses and reserve carbohydrates is, moreover, suggested by the slow rate of adjustment which is characteristic of altered carbohydrate balance relations. The course of respiration in the transition 'is exactly that of an approach to a reversible equilibrium. . . . We picture the opposed processes to be increased production of C from A — B by oxygen activation working against increased consumption of C by glycolysis, which rises with each rise of concentration until the two become adjusted to equality again.' [In Blackman's scheme A represents reserve carbohydrate, B normal (pyranose) hexose and C effective substrate (called heterohexose by Blackman).]

It will be further observed that these considerations suggest that the effective substrate in glycolysis is present in low concentrations since changes in oxygen concentration bring about marked changes in its production and consumption. For this reason a pre-glycolytic reaction consisting merely of the hydrolysis of reserve carbohydrates to normal hexoses is insufficient to meet the case since the normal hexoses are present in relatively high concentration and would therefore undergo slight alterations in concentration.

Such are the considerations which led Blackman to postulate what he called heterohexoses as the direct substrate for glycolysis, and not normal hexoses. We should now have to regard the heterohexoses as phosphorylated hexose.

Blackman's conclusions regarding the course of aerobic respiration subsequent to glycolysis are largely based on Parija's observations, to which reference has been made earlier, on respiration of apples in air and in nitrogen. It will be recalled that when air is replaced by nitrogen the initial rate of respiration in nitrogen at the moment of transference is 1·33 or 1·5 times the rate in air, but this rate slowly falls to a level which may be that of respiration in air, or somewhat higher. This slow fall is that slow transition which, as we have just seen, is characteristic of a change in oxygen concentration, and indicates the pre-glycolytic decrease in production of heterohexoses resulting from the fall in oxygen concentration from 20 per cent. to nothing. At the beginning, therefore, of the nitrogen period, the concentration of glycolytic substrate, which determines the rate of glycolysis, is that obtaining in air. In other words, the initial value of respiration in nitrogen (actually obtained by extrapolation to allow for the diffusion lag as explained in Chapter III) gives a value for glycolysis in air.

Now in anaerobic respiration only one-third of the carbon acted upon in the glycolytic stage finally appears as carbon dioxide; the other two-thirds appear as ethyl alcohol. If, then, the carbon atom is taken as a unit, we can say that respiration in nitrogen, as measured by carbon dioxide evolved, is only one-third of glycolysis, or using Blackman's notation, $Gl=3NR$, where Gl signifies glycolysis and NR nitrogen respiration.

Now we have seen that when air is replaced by nitrogen, before the rate of glycolysis has altered, the respiration increases to 1·5 or 1·33 times that in air. Consequently, using the symbol OR for respiration in air, we have

$$NR = 1·5\ OR\ (\text{or } 1·33\ OR)$$
and
$$Gl = 4·5\ OR\ (\text{or } 4\ OR)$$

This means that under aerobic conditions only one atom of carbon out of every 4 or 4·5 subjected to glycolysis is actually released as carbon dioxide. Consequently for every atom of carbon released as carbon dioxide 3 or 3·5 atoms of carbon are worked back into the system, or, again using Blackman's system of notation, OA (oxidative anabolism) $=3·5\ OR$ (or $3·0\ OR$). It is not clear whether these different ratios (3·5 and 3·0) represent the extremes of a range, or whether the ratios remain constant within a given type. The available evidence appears to point to the latter alternative. If this is so, the lower ratio appears to be associated with late metabolic states of the apple and the higher ratio with earlier ones.

Assuming Blackman's hypothesis, the amount of oxidative anabolism relative to the amount of glycolysis varies in different tissues and probably in the same tissue at different times. With storage tissues and roots values of NR/OR between about 0·5 and 1·4 are common, indicating that from about one-third to three-quarters of the carbon involved in glycolysis is subjected to oxidative anabolism. In some of the young seedlings examined by Leach and Dent the value of $\frac{NR}{OR}$ was found to be one-third or less (cf. Table XV, p. 75), suggesting the absence of oxidative anabolism in such material.

The question naturally arises as to what is the fate of the carbon subjected to oxidative anabolism. While there is at present no very definite answer to this question there are indications that hexose may be resynthesized. Meyerhof demonstrated the production of hexose from pyruvic acid in muscle, while work by Bennet-Clark and Bexon has suggested the possibility of a re-synthesis of hexose in plant tissues. They found that when thin slices of beetroot were placed in sap expressed from such tissue the rate of respiration increased. This appeared to be due to utilization of salts of organic acids present in the sap, for a similar increase in respiration rate resulted from the addition of malate, citrate or succinate of similar concentration (about 0·05N) to that of the acids in the sap. The respiratory quotient rose at the same time from about unity to values between 1·5 and 2·3. If malic or citric acid were completely oxidized to carbon dioxide and water the value would be 1·33, and if succinic acid were the substrate it would be only 1·143. Thus the observed quotients are considerably higher than those which would result from complete oxidation of the acid substrates, and if normal respiration were proceeding at the same time the disparity between the observed and theoretical quotients would be even greater. The high quotients found experimentally would be accounted for if the acids were only partly oxidized to carbon dioxide and water and if part of them was synthesized to hexose. This view is supported by the observation that for every molecule of acid lost only one molecule of carbon dioxide is produced, so that of the four carbon atoms in each molecule of acid lost only one appears in carbon dioxide.

The suggestion was made that two molecules of malic acid, for instance, give rise to one molecule of hexose and two of carbon dioxide. The presence in plants of enzymes effecting the oxidation of malic acid to oxalacetic acid and the decarboxylation of this to pyruvic acid has now been recorded by Vennesland and co-workers. It is thus interesting that before Vennesland's work Bennet-Clark and Bexon had put forward a tentative scheme for the formation of hexose from malic acid through oxalacetic acid and pyruvic acid:

$$\begin{array}{cccc}
\text{COOH} & \text{COOH} & \text{COOH} & \text{CHOH}\text{---} \\
| & | & | & | \\
\text{CHOH} & \text{CO} & \text{CO} & \text{CHOH} \\
| & | & | & | \\
\text{CH}_2 & \text{CH}_2 & \text{CH}_3 & \text{CHOH} \quad \text{O} \\
| & | & & | \\
\text{COOH} & \text{COOH} & [+\text{CO}_2] & \text{CHOH} \\
& & & | \\
& & & \text{CH}\text{---} \\
& & & | \\
& & & \text{CH}_2\text{OH}
\end{array}$$

malic acid   oxalacetic acid   pyruvic acid   hexose

This is not to suggest that the dicarboxylic acids are actual intermediates in respiration although they may play the part assigned to them in the Krebs cycle. Rather it suggests that in normal respiration pyruvic acid forms the starting-point of the anabolic process hypothesized by Blackman.

It may be worthy of note that the organs in which evidence for oxidative anabolism has been produced are mainly those, such as storage organs or senescent fruits, in which active growth is not taking place. In these, if such anabolism does occur, it is reasonable enough to suppose that the product might be hexose, for energy is not required for the building of proteins or other complex substances necessary for growth. In the few seedlings examined there is little or no evidence of oxidative anabolism as the ratio NR/OR is generally not more than 0·33 except in seeds respiring fat, to which the argument based on utilization of carbohydrate does not apply. For the most part evidence relating to other actively growing organs is lacking. However, whatever the ratio NR/OR may be, it is reasonable to suppose that in these there is a linking of anabolic processes with the catabolic ones, so that at least part of the energy released in the degradation of the substrate is utilized in the building up of proteins and other complex substances required for growth and in the maintenance of vital activities such as water and salt absorption, and that intermediates in that degradation may be utilized in the formation of these complex compounds. In the production of proteins, for example, there is reason to believe that the amino-acids, from which the proteins are built, result from the reaction of a substance possessing an $-NH_2$ group, such as ammonia or hydroxylamine, with an α-ketonic acid such as pyruvic acid or oxalacetic acid. We have already seen that pyruvic acid may be the starting-point for the oxidative processes in respiration and that enzyme systems exist which catalyse the production of oxalacetic acid from pyruvic acid, and it is at least possible that pyruvic acid produced in carbohydrate breakdown may form the starting-point for the formation of proteins.

ENERGY TRANSFER IN RESPIRATION

When hexose sugar is burnt with the formation of carbon dioxide and water some 670 kilogram calories are released in the form of heat for each gram-molecule of sugar burnt. In the process of aerobic respiration the end products are the same as those which result from the combustion of hexose and the same amount of energy must have been released from the same amount of sugar. But as pointed out earlier, the conditions of degradation of hexose are very different in the processes of combustion and respiration, and while in the latter a proportion of the energy released is no doubt dissipated in heat, some of it is utilized in the building up of complex substances and in other vital activities.

The evolution of heat by actively respiring tissues such

as germinating seeds and opening flower-buds is readily demonstrated by the use of vacuum flasks. The heat evolved by the respiring tissue is lost very slowly through the walls of the flask and a rise in temperature of the contents of the flask results. The rise of temperature during the development of the inflorescence of *Arum italicum* was recorded as long ago as 1778 by Lamarck. According to a report by Kraus more than 100 years later the temperature of the spadix of this species may rise to $27 \cdot 7°$ C. above that of the surrounding air. This rise in temperature is, however, exceptionally high, and it has been stated that the temperature of an actively growing shoot is not as a rule more than $0 \cdot 3°$ C. above that of the surroundings.

In the economic field it is well known that serious damage is frequently suffered by products such as grain and soya beans, due to their heating as a result of respiratory activity when they are stored in bulk under too moist conditions. It has been adequately proved by Ramstad and Geddes, Snow and Wright, and by Leach, that in these cases the respiratory activity which results in the heating is almost entirely due to contaminating moulds and bacteria. Ramstad and Geddes, by means of an apparatus in which soya beans of various moisture contents were stored under adiabatic conditions, obtained temperature rises of as much as $69 \cdot 4°$ C. in an experimental period of 27 hours.

It is, of course, the fate of the energy derived from the degradation of hexose which is not dissipated as heat which is of particular interest to the student of plant respiration. While it has been recognized for a long time that the catabolic reactions were linked with others whereby energy was transferred and utilized in processes necessary for growth and maintenance, the way in which this transference of energy was effected was quite obscure. Within recent years, however, evidence has been produced indicating that organic phosphates containing energy-rich bonds, particularly the adenosine phosphates

ADP and ATP (cf. p. 102) and enzymes catalysing the transfer of phosphate groupings in these compounds, play an important part in energy transfer.

In the formation of the energy-rich bonds the energy is generally provided by oxidation. In the scheme of hexose breakdown summarized on pages 103 to 109 it will be observed that in reactions 6 and 7 the phosphoglyceraldehyde, after combining with a molecule of phosphoric acid, is oxidized to diphosphoglyceric acid. The energy released in the oxidation is transferred to the bond linking the phosphate to the carboxyl group, which thus becomes an energy-rich bond. In the next reaction this phosphate is transferred to adenosine diphosphate, the energy being transferred along with the phosphate, with the formation of adenosine triphosphate, so that this phosphate group is now attached by an energy-rich bond:

$$\begin{array}{l}CH_2.O.H_2PO_3\\ |\\ CHOH \qquad + A-O-HPO_3 \sim H_2PO_3\\ |\\ CO.O.H_2PO_3\end{array}$$

$$= \begin{array}{l}CH_2.O.H_2PO_3\\ |\\ CHOH \qquad + A-O-HPO_3 \sim HPO_3 \sim H_2PO_3\\ |\\ COOH\end{array}$$

The terminal phosphate grouping of ATP with the energy of the bond linking it to the ADP residue can then be transferred to some other substance. Examples of this have already been cited in dealing with the anaerobic breakdown of sugar. In this way glucose-6-phosphate is formed from glucose and fructose-1,6-diphosphate from fructose-6-phosphate.

Now in the whole course of fermentation or anaerobic respiration it will be observed that for every molecule of hexose utilized two molecules of ATP are dephosphorylated to ADP (in reactions 1 and 3) while four molecules of ADP are phosphorylated to ATP (in reactions 8 and 11, since two molecules of the 3-carbon molecules involved are obtained from each molecule of hexose). There is thus a net gain of two molecules of ATP and of two energy-rich phosphate bonds for every molecule of hexose subjected to glycolysis, and the energy of these bonds has been derived from the hexose. The phosphate is provided by the inorganic phosphate which links up with phosphoglyceraldehyde in reaction 6. In fermentation or anaerobic respiration the energy of two energy-rich phosphate bonds thus become available for transfer for every molecule of hexose utilized. The energy of these two bonds is not much less than the whole of the energy released in the breakdown of a molecule of glucose to carbon dioxide and ethyl alcohol (cf. pp. 66 and 102).

The energy released in the aerobic breakdown of hexose to carbon dioxide and water is very much greater than that released in fermentation or anaerobic respiration, roughly about twenty-five times as much. Most of this will thus be released in the reactions involved in the oxidation phase of aerobic respiration, and it is a reasonable hypothesis that the oxidations occurring in this phase are coupled with phosphorylations. Although pyruvic acid itself is not so phosphorylated there is evidence that in animal tissues oxidation of the various acids of the Krebs cycle, succinic acid, fumaric acid and malic acid, may give rise to phosphorylation of various substances. While practically nothing is known of such a coupling of oxidation with phosphorylation in higher plants, and while, indeed, the existence in plants of a tricarboxylic acid cycle similar to the Krebs cycle is still a matter of hypothesis, it would be premature to pursue the argument further in this place. It is, however, a reasonable hypothesis that the transfer of energy from the hexose forming the respiratory substrate to the more complex substances synthesized as the result of respiratory activity, is brought about through phosphorylations linked with oxidations and the transfer of the energy of energy-rich phosphate bonds. The same means of energy transference may also be involved in the provision of energy for moving salts in opposition to the concentration gradient and for other vital activities. The actual course of transfer of the energy released in the oxidation of hexose to the reactions involved in these vital activities is scarcely more than a matter of conjecture and presents problems which only further experimental research can solve.

LITERATURE CITED IN THE TEXT

AUBERT, E. Recherches sur la respiration et l'assimilation des plantes grasses. *Rev. gén. Bot.*, **4,** 203–82, 321–31, 337–53, 421–41, 497–502, 558–68. 1892

AUDUS, L. J. Mechanical stimulation and respiration rate in the cherry laurel. *New Phyt.*, **34,** 557–80. 1936

— Mechanical stimulation and respiration in the green leaf. II. Investigation on a number of angiospermic species. *New Phyt.*, **38,** 284–8. 1939

— Mechanical stimulation and respiration in the green leaf. III. The effect of stimulation on the rate of fermentation. *New Phyt.*, **39,** 65–74. 1940

— Mechanical stimulation and respiration in the green leaf. Parts IV and V. *New Phyt.*, **40,** 86–95. 1941

BAKER, D., and NELSON, J. M. Tyrosinase and plant respiration. *Journ. Gen. Physiol.*, **26,** 269–76. 1943

BARKER, J. Note on the effect of handling on the respiration of potatoes. *New Phyt.*, **34,** 407–8. 1935

BENNET-CLARK, T. A., and BEXON, D. Water relations of plant cells. III. The respiration of plasmolysed tissues. *New Phyt.*, **42,** 65–92. 1943

BENNETT, E. A note on the presence of pyruvic acid in Ebenezer onions. *Plant Physiol.*, **20,** 461–3. 1945

BIALE, J. B., and YOUNG, R. E. Critical oxygen concentrations for the respiration of lemons. *Amer. J. Bot.*, **34,** 301–9. 1947

BLACKMAN, F. F. The manifestations of the principles of chemical mechanics in the living cell. Presidential Address to Section K (Botany), *Rep. British Ass. Adv. Sci.*, 1908. Publ. London, 1909

— Analytical studies in plant respiration. III. Formulation of a catalytic system for the respiration of apples and its relation to oxygen. *Proc. Roy. Soc.*, B, **103,** 491–523. 1928

— Respiration and Oxygen-Concentration. Fifth International Botanical Congress, Cambridge, 1930. *Report of Proceedings*. Publ. Cambridge, 1931

BLACKMAN, F. F., and PARIJA, P. Analytical studies in plant respiration. I. The respiration of a population of senescent ripening apples. *Proc. Roy. Soc.*, B, **103**, 412–45. 1928

BODNÁR, J. Über die Zymase und Carboxylase der Kartoffel und Zuckerrübe. *Biochem. Zeitschr.*, **73**, 193–210. 1916

Biochemie des Phosphorsäurestoffwechsels der höheren Pflanzen. I. Mitt. Über die enzymatische Überführung der anorganischen Phosphorsäure in organische Form. *Biochem. Zeitschr.*, **165**, 1–15. 1925

BÖHM, J. Ueber die Respiration der Kartoffel. *Bot. Zeit.*, **45**, 671–5, 681–92. 1887

BONNER, J. Biochemical mechanisms in the respiration of the *Avena* coleoptile. *Arch. Biochem.*, **17**, 311–26. 1948

BONNIER, G., and MANGIN, L. Respiration des tissus sans chlorophylle. *Ann. Sci. nat.*, Sér. 6, **18**, 293–379. 1884

BOSWELL, J. G. Oxidation systems in the potato tuber. *Ann. Bot.*, **9**, 55–76. 1945

Metabolic systems in the 'root' of *Brassica napus* L. *Ann. Bot.*, N.S., **14**, 521–43. 1950

BOSWELL, J. G., and WHITING, G. C. A study of the polyphenol oxidase system in potato tubers. *Ann. Bot.*, N.S., **2**, 847–64. 1938

Observations on the anaerobic respiration of potato tubers. *Ann. Bot.*, N.S. **4**, 257–68. 1940

BUCHNER, E. Alkoholische Gärung ohne Hefezellen. *Ber. deut chem. Ges.*, **30**, 117–24, 1110–13. 1897

BUTKEWITSCH, W. Umwandlung der Eiweissstoffe durch die niederen Pilze im Zusammenhange mit einigen Bedingungen ihrer Entwickelung. *Jahr. f. wiss. Bot.*, **38**, 147–240. 1903

CHOUDHURY, J. K. Researches on plant respiration. V. On the respiration of some storage organs in different oxygen concentrations. *Proc. Roy. Soc.*, B, **127**, 238–57. 1939

CHUDIAKOW, N. V. Beiträge zur Kenntnis der intramolekularen Athmung. *Landw. Jahrb.*, **23**, 333–89. 1894

CLAUSEN, H. Beiträge zu Kenntniss der Athmung der Gewächse und des pflanzlichen Stoffwechsels. *Landw. Jahrb.*, **19**, 893–930. 1890

CONN, E., VENNESLAND, B., and KRAEMER, L. M. Distribution of a triphosphopyridine nucleotide-specific enzyme catalysing the reversible oxidative decarboxylation of malic acid in higher plants. *Arch. Biochem.*, **23**, 179–197. 1949

CRUICKSHANK, W. Some observations of the nature of sugar, etc. In 'An Account of two Cases of Diabetes mellitus', by John Rollo, Vol. II. London, 1797

DAMODARAN, M., and NAIR, K. R. Glutamic acid dehydrogenase from germinating seeds. *Biochem. Journ.*, **32**, 1064–74. 1938

DE BOER, S. R. Respiration of phycomyces. *Rec. trav. bot. Néerlandais*, **25**, 117–240. 1928

DENNY, F. E. Respiration of Gladiolus corms during prolonged dormancy. *Contrib. Boyce Thompson Inst.*, **10**, 453–60. 1939

Accumulation of carbon dioxide in potato tuber tissue under conditions for the continuous removal of the exhaled gas. *Contrib. Boyce Thompson Inst.*, **14**, 315–322. 1946

Changes in oxygen, carbon dioxide, and pressure caused by plant tissue in a closed space. *Contrib. Boyce Thompson Inst.*, **14**, 383–96. 1947

Respiration rate of plant tissue under conditions for the progressive partial depletion of the oxygen supply. *Contrib. Boyce Thompson Inst.*, **14**, 419–42. 1947

Effect upon plant respiration caused by changes in the oxygen concentration in the range immediately below that of normal air. *Contrib. Boyce Thompson Inst.*, **15**, 61–70. 1948

DETMER, W. Vergleichende Physiologie des Keimungsprocesses der Samen. Jena, 1880

DEVAUX, H. Porosité du fruit des Cucurbitacées. *Rev. gén. bot.*, **3**, 49–56. 1891

Asphyxie spontanée et production d'alcool dans les tissus profonds des tiges ligneuses poussant dans les conditions naturelles. *Comp. rend.*, **128**, 1346–9. 1899

DIAKONOW, N.W. Intramolekulare Athmung und Gährthätigkeit der Schimmelpilze. *Ber. deut. bot. Ges.*, **4**, 2–7. 1886

Ueber die sogenannte intramolekulare Athmung der Pflanzen. *Ber. deut. bot. Ges.*, **4**, 411–13. 1886

DIXON, M. Multi-enzyme systems. Cambridge, 1949

EMBDEN, G., and DEUTIGE, H. J. Über die Bedeutung der

Phosphoglycerinsäure für die Glycolyse in der Muskulatur. *Zeitschr. f. physiol. Chem.*, **230**, 24-49. 1934

EMBDEN, G., DEUTIGE, H. J., and KRAFT, G. Über die intermediären Vorgänge bei der Glykolyse in der Muskulatur. *Klin. Wochenschr.*, **12**, 213-15. 1933

Über das Vorkommen einer optisch-aktiven Phosphoglycerinsäure bei den Glykolyse in der Muskulatur. *Zeitschr. f. physiol. Chem.*, **230**, 12-28. 1934

EMBDEN, G., and JOST, H. Über die Zwischenstufen der Glykolyse in der quergestreiften Muskulatur. *Zeitschr. f. physiol. Chem.*, **230**, 68-89. 1934

FERNANDES, D. S. Aerobe und anaerobe Atmung bei Keimlingen von *Pisum sativum*. *Rec. trav. bot. Néerlandais*, **20**, 107-256. 1923

FIDLER, J. C. The conserving influence of oxygen on respirable substrate. *Ann. Bot.*, N.S. **12**, 421-6. 1948

GARREAU. De la respiration chez les plantes (1). *Ann. sci. nat., Bot.*, 3ᵉ Sér., **15**, 5-36. 1851

Nouvelles recherches sur la respiration des plantes. *Ann. sci. nat., Bot.*, 3ᵉ Sér., **16**, 271-92. 1851

GENEVOIS, L. Über Atmung und Gärung in grünen Pflanzen. II. *Biochem. Zeitschr.*, **191**, 147-57. 1927

GERBER, C. Etude comparée de la respiration des graines oléagineuses pendent leur développement et pendant leur germination—Relations entre cette respiration et les réactions chimiques dont la graine est la siège. *Actes du Congrès Internat. de Botanique*, 59-101. Paris, 1900

GERHART, A. R. Respiration in strawberry fruits. *Bot. Gaz.*, **89**, 40-66. 1930

GODDARD, D. R., and MEEUSE, B. J. D. Respiration of higher plants. *Ann. Rev. Plant Physiol.*, **1**, 207-32. 1950

GODLEWSKI, E., and POLZENIUSZ, F. Ueber die intramolekulare Athmung von in Wasser gebrachten Samen und über die dabei stattfindende Alkoholbildung. *Ann. Akad. Wiss. Krakau*, 1901. Reference in *Bot. Centralblatt*, **89**, 713. 1902

GUSTAFSON, F. G. Growth studies on fruits. Respiration of tomato fruits. *Plant Physiol.*, **4**, 349-56. 1929

Production of alcohol and acetaldehyde by tomatoes. *Plant Physiol.*, **9**, 359-67. 1934

Äthylalkohol und Acetaldehyd in gewissen Arten von Kakteen. *Biochem. Zeitschr.*, **272**, 172–9. 1934

HANES, C. S. The breakdown and synthesis of starch by an enzyme system from pea seeds. *Proc. Roy. Soc.*, B, **128**, 421–50. 1940

The reversible formation of starch from glucose-1-phosphate catalysed by potato phosphorylase. *Proc. Roy. Soc.*, B, **129**, 174–208. 1940

HANES, C. S., and BARKER, J. The physiological action of cyanide. I. The effects of cyanide on the respiration and sugar content of the potato at 15° C. *Proc. Roy. Soc.*, B, **108**, 95–118. 1931

HARDEN, A., and YOUNG, W. J. The alcoholic ferment of yeast-juice. *Proc. Physiol. Soc.*, i–ii, Nov. 12, 1904, in *Journ. Physiol.*, **32**, 1904–5

The function of phosphates in the fermentation of glucose by yeast-juice. *Proc. Roy. Soc.*, B, **80**, 299–311. 1908

HARRINGTON, G. T. Respiration of apple seeds. *Journ. Agric. Res.*, **23**, 117–30. 1923

HAWORTH, W. N. The constitution of sugars. London, 1929

HILL, G. R. Respiration of fruits and growing plant tissues in certain gases, with reference to ventilation and fruit storage. *Cornell Univ. Agric. Exper. Sta. Bull.*, 330. 1913

HOPKINS, E. F. Variation in sugar content in potato tubers caused by wounding and its possible relation to respiration. *Bot. Gaz.*, **84**, 75–88. 1927

HOVER, J. M., and GUSTAFSON, F. G. Rate of respiration as related to age. *Journ. Gen. Physiol.*, **10**, 33–9. 1926

INAMDAR, R. S., and SINGH, B. N. Studies in the respiration of tropical plants. I. Seasonal variations in aerobic and anaerobic respiration in the leaves of *Artocarpus integrifolia*. *Journ. Indian Bot. Soc.*, **6**, 133–212. 1927

INGEN-HOUSZ, J. Experiments upon vegetables, discovering their great power of purifying the common air in the sunshine and of injuring it in the shade and at night; to which is joined a new method of examining the accurate degree of salubrity of the atmosphere. London, 1779

IRVING, ANNIE A. The effect of chloroform upon respiration and assimilation. *Ann. Bot.*, **25**, 1077–99. 1911

IVANOV, L. (IWANOFF, L.). Zur Frage nach der Beteiligung der Zwischenprodukte der alkoholischen Gärung an der Sauerstoffatmung. *Ber. deut. bot. Ges.*, **32,** 191–6. 1914

IVANOV, N. (IWANOFF, N.). Die Wirkung der nützlichen und schädlichen Stimulatoren auf die Atmung der lebenden und abgetoteten Pflanzen. *Biochem. Zeitschr.*, **32,** 74–96. 1911

JAMES, G. M., and JAMES, W. O. The formation of pyruvic acid in barley respiration. *New Phyt.*, **39,** 266–70. 1940

JAMES, W. O., and ARNEY, S. E. Phosphorylation and respiration in barley. *New Phyt.*, **38,** 340–51. 1939

JAMES, W. O., and BEEVERS, H. The respiration of *Arum* spadix. A rapid respiration, resistant to cyanide. *New Phyt.*, **49,** 353–74. 1950

JAMES, W. O., HEARD, C. R. C., and JAMES, G. M. On the oxidative decomposition of hexosediphosphate by barley. The role of ascorbic acid. *New Phyt.*, **43,** 62–74. 1944

JAMES, W. O., JAMES, G. M., and BUNTING, A. H. On the method of formation of pyruvic acid by barley. *Biochem. Journ.*, **35,** 588–94. 1939

JAMES, W. O., and NORVAL, I. F. The respiratory decomposition of pyruvic acid by barley. *New Phyt.*, **37,** 455–73. 1938

JENSEN, P. BOYSEN. Studien über den genetischen Zusammenhang zwischen der normalen und intramolekularen Atmung der Pflanzen. *Kgl. Danske Videnskabernes Selskab. Biol. Med.*, IV, **1,** 34 pp. 1923

The connection between the oxybiotic and anoxybiotic respiration in plants. Fifth International Botanical Congress, Cambridge, 1930. *Report of Proceedings*, 421–2. Publ. Cambridge, 1931

JÖNSSON, B. Recherches sur la respiration et l'assimilation des Muscinées. *Comp. rend.*, **119,** 440–3. 1894

KARLSEN, A. Comparative studies on respiration. XXVIII. The effect of anesthetics on the production of carbon dioxide by wheat under aërobic and anaërobic conditions. *Amer. Journ. Bot.*, **12,** 619–24. 1925

KEILIN, D. Cytochrome and respiratory enzymes. *Proc. Roy. Soc.*, B, **104,** 206–52. 1929

KEILIN, D., and HARTREE, E. F. Properties of catalase. Catalysis of coupled oxidation of alcohols. *Biochem. Journ.*, **39**, 293–301. 1945

KERMACK, W. O. [Recent Advances in] Biochemistry. Some reactions and enzymes concerned in glycolysis. *Sci. Prog.*, **37**, 283–93. 1949

KIDD, F. The controlling influence of carbon dioxide in the maturation, dormancy, and germination of seeds. Part I. *Proc. Roy. Soc.*, B, **87**, 408–21. 1914

The controlling influence of carbon dioxide in the maturation, dormancy, and germination of seeds. Part II. *Proc. Roy. Soc.*, B, **87**, 609–25. 1914

The controlling influence of carbon dioxide. Part III. The retarding effect of carbon dioxide on respiration. *Proc. Roy. Soc.*, B, **89**, 136–56. 1915

KIDD, F., and WEST, C. Physiology of fruit. I. Changes in the respiratory activity of apples during their senescence at different temperatures. *Proc. Roy. Soc.*, B, **106**, 93–109. 1930

KIDD, F., WEST, C., and BRIGGS, G. E. A quantitative analysis of the growth of *Helianthus annuus*. Part I. The respiration of the plant and of its parts throughout the life cycle. *Proc. Roy. Soc.*, B, **92**, 368–84. 1921

KIDD, F., WEST, C., GRIFFITHS, D. G., and POTTER, N. A. An investigation on the changes in chemical composition and respiration during the ripening and storage of Conference pears. *Ann. Bot.*, N.S., **4**, 1–30. 1940

KLEIN, G., and PIRSCHLE, K. Acetaldehyd als Zwischenprodukt der Pflanzenatmung. *Biochem. Zeitschr.*, **168**, 340–60. 1926

Quantitative Untersuchungen über die Verwertbarkeit verschiedener Stoffe für die Pflanzenatmung. *Biochem. Zeitschr.*, **176**, 20–31. 1926

KOBEL, M., and SCHEUER, M. Über den Kohlenhydratumsatz im Tabakblatt. Nachweis von Methylglyoxal als Zwischenprodukt im Stoffwechsel grüner Blätter. *Biochem. Zeitschr.*, **216**, 216–23. 1929

KOSINSKI, I. Die Athmung bei Hungerzuständen und unter Einwirkung von mechanischen und chemischen Reizmitteln bei *Aspergillus niger*. *Jahrb. f. wiss. Bot.*, **37**, 137–204. 1902

KOSTYCHEV, S. (KOSTYTSCHEW, S.). Der Einfluss des

Substrates auf die anaërobe Athmung der Schimmelpilze. *Ber. deut. bot. Ges.*, **20**, 327–34. 1902

Über die normale und die anaërobe Athmung bei Abwesenheit von Zucker. *Jahrb. f. wiss. Bot.*, **40**, 563–92. 1904

Über die Alkoholgärung von *Aspergillus niger*. *Ber. deut. bot. Ges.*, **25**, 44–50. 1907

Über Alkoholgärung. I. Über die Bildung von Acetaldehyd bei der alkoholischen Zuckergärung. *Zeitschr. physiol. Chem.*, **79**, 130–45. 1912

Plant Respiration. Authorized edition in English with editorial notes. Translated and edited by Charles J. Lyon. Philadelphia, 1927

KOSTYCHEV, S. (KOSTYTSCHEW, S.), HÜBBENET, E., and SCHELOUMOFF, A. Über die Bildung von Acetaldehyd bei den anaeroben Atmung der Pappelblüten. *Zeitschr. f. physiol. Chem.*, **83**, 105–11. 1913

KREBS, H. A., and EGGLESTON, L. V. The oxidation of pyruvate in pigeon breast muscle. *Biochem. Journ.*, **34**, 293–301. 1940

LEACH, W. Researches on plant respiration. IV. The relation between the respiration in air and in nitrogen of certain seeds during germination. (b) Seeds in which carbohydrates constitute the chief food reserve. *Proc. Roy. Soc.*, B, **119**, 507–21. 1936

Studies on the metabolism of cereal grains. III. The influence of atmospheric humidity and mould infection on the carbon dioxide output of wheat. *Canadian Journ. Res.*, C, **22**, 150–61. 1944

LEACH, W., and DENT, K. W. Researches on plant respiration. III. The relation between the respiration in air and in nitrogen of certain seeds during germination. (a) Seeds in which fats constitute the chief food reserve. *Proc. Roy. Soc.*, B, **116**, 150–69. 1934

LECHARTIER, G., and BELLAMY, F. Étude sur les gaz produits par les fruits. *Comp. rend.*, **69**, 356–60. 1869

De la fermentation des fruits. *Comp. rend.*, **69**, 466–9. 1869

De la fermentation des fruits. *Comp. rend.*, **75**, 1203–6. 1872

De la fermentation des pommes et des poires. *Comp. rend.*, **79**, 949–52. 1874

LEVY, H., and SCHADE, A. L., with the technical assistance of L. BERGMANN and S. HARRIS. Studies in the respiration of the white potato. II. Terminal oxidase system of potato tuber respiration. *Arch. Biochem.*, **19**, 273–86. 1948

LIPMANN, F. Metabolic generation and utilization of phosphate bond energy. *Advances in Enzymology*, **1**, 99–162. 1941

LIVINGSTON, E., and FRANCK, J. Assimilation and respiration of excised leaves at high concentrations of carbon dioxide. *Amer. J. Bot.*, **27**, 449–58. 1940

LOHMANN, K. Konstitution der Adenylpyrophosphorsäure und Adenosindiphosphorsäure. *Biochem. Zeitschr.*, **282**, 120–3. 1935

LUNDEGÅRDH, H., and BURSTRÖM, H. Untersuchungen über die Salzaufnahme der Pflanze. III. Quantitative Bezeichungen zwischen Atmung und Anionaufnahme. *Biochem. Zeitschr.*, **261**, 235–51. 1933

LUNDSGAARD, E. Die Monojodessigsäurewirkung auf die enzymatische Kohlenhydratspaltung. *Biochem. Zeitschr.*, **220**, 1–7. 1930

LYON, C. J. The effect of phosphates on respiration. *Journ. Gen. Physiol.*, **6**, 299–306. 1924.

MAGNESS, J. R. Composition of gases in intercellular spaces of apples and potatoes. *Bot. Gaz.*, **70**, 308–16. 1920

MAIGE, A., and NICOLAS, G. Recherches sur l'influence des solutions sucrées de divers degrés de concentration sur la respiration, la turgescence et la croissance de la cellule. *Ann. sci. nat., Bot.*, Sér. 9, **12**, 315–68. 1910

MAIGE, G. Recherches sur la respiration des différentes pièces florales. *Ann. sci. nat., Bot.*, Sér. 9, **14**, 1–62. 1911

MALPIGHI, M. Anatomes plantarum pars altera. Londini, 1679.

MAQUENNE, L. Sur la respiration des feuilles. *Comp. rend.*, **119**, 100–2. 1894

Sur le mécanisme de la respiration végétale. *Comp. rend.*, **119**, 697–9. 1894

MAQUENNE, L., and DEMOUSSY, E. Sur la valeur des coefficients chlorophylliens et leur rapports avec les quotients respiratoires réels. *Comp. rend.*, **156**, 506–12. 1913

MARSH, P. B., and GODDARD, D. R. Respiration and fermentation in the carrot, *Daucus Carota*. I. Respiration. *Amer. J. Bot.*, **26**, 724–8. 1939

Respiration and fermentation in the carrot, *Daucus Carota*. II. Fermentation and the Pasteur effect. *Amer. J. Bot.*, **26**, 767–72. 1939

MATTHAEI, GABRIELLE L. C. Experimental Researches on vegetable assimilation and respiration. III. On the effect of temperature on carbon dioxide assimilation. *Phil. Trans. Roy. Soc. London* B, **197**, 47–105. 1904

MAYER, A. Über den Verlauf der Athmung beim keimenden Weizen. *Die landw. Versuchs-Stationen*, **18**, 245–279. 1875

MEYERHOF, O. Über den Einfluss des Sauerstoffs auf die alkoholische Gärung der Hefe. *Biochem. Zeitschr.*, **162**, 43–86. 1925

MEYERHOF, O., and KIESSLING, W. Über das Auftreten und den Umsatz der α-Glycerinphophorsäure bei der enzymatischen Kohlenhydratspaltung. *Biochem. Zeitschr.*, **264**, 40–71. 1933

Über die phosphorylierten Zwischenprodukte und die letzten Phasen der alkoholischen Gärung. *Biochem. Zeitschr.*, **267**, 313–48. 1933

MEYERHOF, O., LOHMANN, K., and MEIER, R. Über die Synthese des Kohlenhydrats im Muskel. *Biochem. Zeitschr.*, **157**, 459–91. 1925

MEYERHOF, O., and MCEACHERN, D. Über anaerobe Bildung und Schwund von Brenztraubensäure in der Muskulatur. *Biochem. Zeitschr.*, **260**, 417–45. 1933

MOHL, H. VON. Grundzüge der Anatomie und Physiologie der vegetabilischen Zelle. Braunsweig, 1851

MORGAN, E. J. Pyruvic acid in the juice of the onion (*Allium Cepa*). *Nature*, **157**, 512. 1946

MÜLLER, D. Ein neues Enzym—Glycoseoxydase—aus *Aspergillus niger*. Den Kgl. *Veterinær- og Landbohojskole Aarskrift*, 329–31. 1925

Studien über ein neues Enzym Glykoseoxydase. I. *Biochem. Zeitschr.*, **199**, 136–70. 1928

MÜLLER-THURGAU, H., and SCHNEIDER-ORELLI, O. Beiträge zur Kenntnis der Lebensvorgänge in ruhenden Pflanzenteilen. *Flora*, **101**, 309–72. 1910

NABOKICH, A. J. Über anaeroben Stoffwechsel von Samen

in Saltpeterlösungen. *Ber. deut. bot. Ges.*, **21**, 398–403. 1903

Über die intramolekulare Atmung der höheren Pflanzen. *Ber. deut. bot. Ges.*, **21**, 467–76. 1903

NÄGELI, C. Theorie der Gärung. München, 1879

NANCE, J. P. (see also PHILLIPS, J. W.). A comparison of carbohydrate loss and carbon dioxide production during fermentation by barley roots. *Amer. Journ. Bot.*, **36**, 274–6. 1949

NĚMEC, A., and DUCHOŇ, F. Versuche über Vorkommen und Wirkung der Saccharophosphatase im Pflanzenorganismus. *Biochem. Zeitschr.*, **119–20**, 73–80. 1921

NEUBERG, C. Von der Chemie der Gärungserscheinungen. *Ber. deut. chem. Ges.*, **55**, 3624–38. 1922

NEUBERG, C., and COHEN, CLARA. Über die Bildung von Acetaldehyd und die Verwirklichung der zweiten Vergärungsform bei verschiedenen Pilzen. *Biochem. Zeitschr.*, **122**, 204–24. 1921

NEUBERG, C., and GORR, G. Über die Mechanismus der Milchsäurebildung bei Phanerogamen. *Biochem. Zeitschr.*, **171**, 475–84. 1926

NEUBERG, C., and GOTTSCHALK, A. Beobachtungen über den Verlauf der anaeroben Pflanzenatmung. *Biochem. Zeitschr.*, **151**, 167–8. 1924

Über den Nachweis von Acetaldehyd als Zwischenstufe bei der anaeroben Atmung höherer Pflanzen. *Biochem. Zeitschr.*, **160**, 256–60. 1925

NEUBERG, C., and KARCZAG, L. Über zuckerfreie Hefegärungen. IV. Carboxylase, ein neues Enzym der Hefe. *Biochem. Zeitschr.*, **36**, 68–75. 1911

Über zuckerfreie Hefegärungen. V. Zur Kenntnis der Carboxylase. *Biochem. Zeitschr.*, **36**, 76–81, 1911

NEUBERG, C., and KOBEL, M. Umwandlung von Phosphoglycerinsäure durch die Fermente gekeimter Erbsen und Bohnen. *Biochem. Zeitschr.*, **272**, 457–8. 1934

NEUBERG, C., and REINFURTH, ELSA. Die Festlegung der Aldehydstufe bei der alkoholischen Gärung. Ein experimenteller Beweis der Acetaldehyd-Brenztraubensäuretheorie. *Biochem. Zeitschr.*, **89**, 365–414. 1918

NICOLAS, M. G. Contribution à l'étude des variations de la respiration des végétaux avec l'âge. *Rev. gén. Bot.*, **30**, 209–25. 1918

Anthocyane et échanges respiratoires des feuilles. *Comp. rend.*, **167**, 130–3. 1918

OLNEY, A. J. Temperature and respiration of ripening bananas. *Bot. Gaz.*, **82**, 415–26. 1926

ONSLOW, MURIEL WHELDALE. The principles of plant biochemistry. Part I. Cambridge, 1931

ONSLOW, MURIEL WHELDALE, and ROBINSON, MURIEL E. R. Oxidizing enzymes. IX. On the Mechanism of Plant Oxidases. *Biochem. Journ.*, **20**, 1138–45. 1926

PALLADIN, W. (PALLADINE, W.). Recherches sur la correlation entre la respiration des plantes et les substances azotées actives. *Rev. gén. Bot.*, **8**, 225–48. 1896

Ueber normale und intramolekulare Atmung der einzelligen Alge *Chlorothecium saccharophilum*. *Centralbl. f. Bakt.*, 2 Abt., **11**, 146–53. 1903

Das Blut der Pflanzen. *Ber. deut. bot. Ges.*, **26a**, 125–32. 1908

Die Verbreitung der Atmungschromogene bei den Pflanzen. *Ber. deut. bot. Ges.*, **26a**, 378–89. 1908

Über die Bildung der Atmungschromogene in den Pflanzen. *Ber. deut. bot. Ges.*, **26a**, 389–94. 1908

PARIJA. P. Analytic studies in plant respiration. II. The respiration of apples in nitrogen and its relation to respiration in air. *Proc. Roy. Soc.*, B, **103**, 446–90. 1928

PASTEUR, L. Faits nouveaux pour servir à la connaissance de la théorie des fermentations proprement dites. *Comp. rend.*, **75**, 784–90. 1872

PFEFFER, W. Das Wesen und die Bedeutung der Athmung in der Pflanze. *Landw. Jahrb.*, **7**, 805–34. 1878

Über intramolekulare Athmung. *Untersuchungen aus dem bot. Inst. zu Tübingen*, **1**, 636–85. 1885

PFLÜGER, E. Beiträge zur Lehre von der Respiration. I. Ueber die physiologische Verbrennung in den lebendigen Organismen. *Arch. f. d. ges. Physiol.*, **10**, 251–367. 1875

PHILLIPS, J. W. (see also NANCE, J. P.). Studies on fermentation in rice and barley. *Amer. Journ. Bot.*, **34**, 62–72. 1947

PLATENIUS, H. Effect of temperature on the respiration rate and the respiratory quotient of some vegetables. *Plant Physiol.*, **17**, 179–97. 1942

POLOVZOV, V. Researches on plant respiration. (In Russian.) 1901. (Cited on the authority of Kostychev, 1927)

PRATT, H. K., and BIALE, J. B. Relation of the production of an active emanation to respiration in the avocado fruit. *Plant Physiol.*, **19**, 519–28. 1944

PURIEWITSCH, K. Physiologische Untersuchungen über Pflanzenathmung. *Jahrb. f. wiss. Bot.*, **35**, 573–610. 1900

RAMSTAD, P. E., and GEDDES, W. F. The respiration and storage behaviour of soybeans. *Univ. Minnesota Agric. Exper. Sta., Tech. Bull.*, **156**, 54 pp., 1942

RICHARDS, H. M. The respiration of wounded plants. *Ann. Bot.*, **10**, 531–82. 1896

RISCHAVI, L. Zur Frage über die Athmung der Pflanzen. *Schriften der neurussischen Ges. der Naturforscher*, **5**, 50 pp. 1877. (In Russian. Abstract in *Just's Botan. Jahresber.*, p. 271. 1879)

ROBERTSON, R. N. Studies in the metabolism of plant cells. I. Accumulation of chlorides by plant cells and its relation to respiration. *Australian J. Exp. Biol. and Med. Sci.*, **19**, 265–78. 1941

ROUX, E. R. Respiration and maturity in peaches and plums. *Ann. Bot.*, N.S., **4**, 317–27. 1940

SACHS, J. Vorlesungen über Pflanzenphysiologie. Leipzig, 1882. (English Translation by H. M. Ward: Lectures on the Physiology of Plants. Oxford, 1887)

SAUSSURE, T. DE. La formation de l'acide carbonique est-elle essentielle à la végétation? *Ann. de Chim.*, **24**, 135–49. 1797

Recherches chimiques sur la végétation. Paris, 1804

De l'action des fleurs sur l'air, et de leur chaleur propre. *Ann. de Chim. et de Phys.*, **21**, 279–303. 1822

SCHADE, A. L., BERGMANN, L., and BYER, A. Studies on the respiration of the white potato. I. Preliminary investigation of the endogenous respiration of potato slices and catechol oxidase activity. *Arch. Biochem.*, **18**, 85–96. 1948

SCHADE, A. L., and LEVY, H. Studies on the respiration of the white potato. III. Changes in the terminal oxidase pattern of potato tissue associated with time of suspension in water. *Arch. Biochem.*, **20**, 211–19. 1949

SCHEELE, C. W. Chemische Abhandlungen von Luft und Feuer. 1777

SEIFRIZ, W. Anaerobic respiration. *Science*, **101**, 88–9. 1945

SENEBIER, J. Physiologie végétale, contenant une description des Organes des Plantes, et une exposition des Phénomènes produits par leur organisation. 5 vols. Genève, 1800

SNOW, D., and WRIGHT, N. C. The respiration rate and loss of dry matter from stored bran. *Journ. Agric. Sci.*, **35**, 126–32. 1945

STENLID, G. Some notes on the effect of sodium azide, 2,4-dinitrophenol, and ortho-phenanthroline upon oxygen consumption in green leaves. *Physiol. Plant.*, **2**, 61–9. 1949

STEWARD, F. C., and PRESTON, C. The effect of salt concentration upon the metabolism of potato discs and the contrasted effect of potassium and calcium salts which have a common ion. *Plant Physiol.*, **16**, 85–116. 1941

STICH, CONRAD. Die Athmung der Pflanzen bei verminderter Sauerstoffspannung und bei Verletzungen. *Flora* (N.R., **49**), **74**, 1–57. 1891

STILES, W., and DENT, K. W. Researches on plant respiration. VI. The respiration in air and in nitrogen of thin slices of storage tissues. *Ann. Bot.*, N.S. **11**, 1–34. 1947

STILES, W., and LEACH, W. On the use of the katharometer for the measurement of respiration. *Ann. Bot.*, **45**, 461–88. 1931

Researches on plant respiration. II. Variations in the respiratory quotient during germination. *Proc. Roy. Soc.*, B, **113**, 405–28. 1933

STOKLASA, J., and CZERNY, F. Isolierung des die anaerobe Atmung der Zelle der hoher organisierten Pflanzen und Thiere bewirkenden Enzyms. *Ber. deut. chem. Ges.*, **36**, 622–34. 1903

TANKÓ, B. Hexosephosphates produced by higher plants. *Biochem. Journ.*, **30**, 692–700. 1936

TAYLOR, D. L. Influence of oxygen tension on respiration, fermentation and growth in wheat and rice. *Amer. J. Bot.*, **29**, 721–38. 1942

THOMAS, M. The controlling influence of carbon dioxide. V. A quantitative study of the production of ethyl alcohol and acetaldehyde by cells of the higher plants in relation to concentration of oxygen and carbon dioxide. *Biochem. Journ.*, **19**, 927–47. 1925

The production of ethyl alcohol and acetaldehyde by apples in relation to the injuries occurring in storage. Part I. Injuries to apples occurring in the absence of oxygen and in certain mixtures of carbon dioxide and oxygen. *Ann. App. Biol.*, **16**, 444–57. 1929

THOMAS, M., and FIDLER, J. C. Studies in zymasis. VI. Zymasis by apples in relation to oxygen concentration. *Biochem. Journ.*, **27**, 1629–42. 1933

Studies in zymasis. VIII. The discovery and investigation of aerobic HCN zymasis in apples treated with hydrogen cyanide; and comparisons with other forms of zymasis. *New Phyt.*, **40**, 217–39. 1941

Studies in zymasis. IX. The influence of HCN on the respiration of apples, and some evaluations of the 'Pasteur effect'. *New Phyt.*, **40**, 240–61. 1941

THORNTON, N. C. Oxygen regulates the dormancy of the potato. *Contrib. Boyce Thompson Inst.*, **10**, 339–61. 1939

THORNTON, N. C., and DENNY, F. E. Oxygen intake and carbon dioxide output of Gladiolus corms after storage under conditions which prolong the rest period. *Contrib. Boyce Thompson Inst.*, **11**, 421–30. 1941

TURNER, J. S. On the relation between respiration and fermentation in yeast and the higher plants. A review of our knowledge of the effect of iodoacetate on the metabolism of plants. *New Phyt.*, **36**, 142–69. 1938

The respiratory metabolism of carrot tissue. I. Material and methods. *New Phyt.*, **37**, 232–53. 1938

The respiratory metabolism of carrot tissue. II. The effect of sodium monoiodoacetate on the respiration and fermentation. *New Phyt.*, **37**, 289–311. 1938

The respiratory metabolism of carrot tissue. III and IV. Part III. The drift of respiration and fermentation in tissue slices, with notes on the respiratory quotient. Part IV. Oxidative anabolism. *Australian Journ. Exp. Biol. and Med. Sci.*, **18**, 275–306. 1940

VENNESLAND, B. The β-carboxylases of plants. II. The distribution of oxalacetic carboxylase in plant tissues. *Journ. Biol. Chem.*, **178**, 591–7. 1949

VENNESLAND, B., GOLLUB, M. C., and SPECK, J. F. The β-carboxylases of plants. I. Some properties of oxalacetic carboxylase and its quantitative assay. *Journ. Biol. Chem.*, **178**, 301–14. 1949

VICKERY, H. B., PUCHER, G. W., WAKEMAN, A. J., and LEAVENWORTH, C. S. The metabolism of amides in green plants. I. The amides of the tobacco leaf. *Journ. Biol. Chem.*, **119**, 369–82. 1937

VIGNOL, M. Contribution à l'étude des Bacteriacées. 1889. (Cited from Kostychev, 1927)

VIRTANEN, A. I., and LAINE, I. Investigations on the root nodule bacteria of leguminous plants. XXII. The excretion products of root nodules. The mechanism of N-fixation. *Biochem. Journ.*, **33**, 412–27. 1939

WARBURG, O. Über die Grundlagen der Wielandschen Atmungstheorie. *Biochem. Zeitschr.*, **141**, 518–23. 1923

WARDLAW, C. W., and LEONARD, E. R. Studies in tropical fruits. IX. The respiration of bananas during ripening at tropical temperatures. *Ann. Bot.*, N.S., **4**, 269–315. 1940

WEHNER, O. Untersuchungen über die chemische Beinflussbarkeit des Assimilationsapparates. *Planta*, **6**, 543–90. 1928

WIELAND, H. Über den Verlauf der Oxydationsvorgänge. *Ber. deut. chem. Ges.*, **55**, 3639–48. 1923

WORTMANN, J. Ueber die Beziehungen der intramolecularen zur normalen Athmung der Pflanzen. *Arbeiten des bot. Inst. in Würzburg*, **2**, 500–20. 1880

YEMM, E. W. The respiration of barley plants. II. Carbohydrate concentration, and carbon dioxide production in starving leaves. *Proc. Roy. Soc.*, B, **117**, 504–525. 1935

Respiration of barley plants. III. Protein catabolism in starving leaves. *Proc. Roy. Soc.*, B, **123**, 243–73. 1937

Respiration of barley plants. IV. Protein metabolism and the formation of amides in starving leaves. *Proc. Roy Soc.*, B, **136**, 632–49. 1950

ZALESKI, W. Über die Rolle Reduktionsprozesse bei der Atmung der Pflanzen. *Ber. deut. bot. Ges.*, **28**, 319–29. 1910

Zum Studium der Atmungsenzyme der Pflanzen. *Biochem. Zeitschr.*, **31**, 195–214. 1911

Über die Verbreitung der Carboxylase in den Pflanzen. *Ber. deut. bot. Ges.*, **31**, 349–53. 1913

Beiträge zur Kenntnis der Pflanzenatmung. *Ber. deut. bot. Ges.*, **31**, 354–61. 1913

ZALESKI, W., and MARX, E. Zur Frage der Wirkung der Phosphate auf die postmortale Atmung der Pflanzen. *Biochem. Zeitschr.*, **43**, 1–6. 1912

Über die Carboxylase bei höheren Pflanzen. *Biochem. Zeitschr.*, **47**, 184–5. 1912

Über die Rolle der Carboxylase in den Pflanzen. *Biochem. Zeitschr.*, **48**, 175–80. 1913

INDEX

Abies excelsa, anaerobic respiration of, 69
Abrahams, M. D., 123
acetaldehyde in fruits, 78–80
— production in fermentation and respiration, 64–5, 78, 83, 87, 100–1, 107–8, 111, 118–19, 130
acetic acid in anthocyanin-containing leaves, 27
acids, effect of, on respiration, 55, 136
—, organic, and respiration, 26–28, 48, 65, 131, 136
Acokanthera spectabilis, respiratory quotients of green and red leaves of, 27
aconitase, 119, 123
'active' fructose, 40
adenosine, 101
— diphosphate, 102–3, 106–7, 139–41
— monophosphate, 102, 110
— triphosphate, 101–3, 106–7, 139–41
adenylic acid, 102, 110
ADP, see adenosine diphosphate
aerobic and anaerobic respiration, connexion between, 82, 84–9, 111
— respiration, 5, 9, 13–58, 80, 82, 98, 118–42
Afanassjewa, M., 64
age, respiration intensity and, 37–8
Ailanthus, respiratory quotient of leaves of, 17
alcohol, production of, in aerobic respiration, 60–6, 76, 78, 84–5, 130–1
— dehydrogenase, 108–9, 112
— in fruits, 78–80
alcoholic fermentation, 5, 60–1, 82–4, 87, 97–109
aldehyde mutase, 108
aldolase, 104
aldoses, constitution of, 92–5
allicine, 111
alliinase, 111

alliine, 111
Allium Cepa, 111
amino-acids, aerobic breakdown of, 91
ammonia production in respiration, 22–3, 91
ammonium salts, utilization of, in respiration, 6
amylase, 90
anaerobic respiration, 5–6, 49, 51, 59–80, 82–9, 108–9, 122, 130–6, 140–1
— zymasis, 78–80
Andersson, B., 120
androecium, respiration intensity of, 33–4
aneurin, 107
anion respiration, 54–5
anthocyanin, respiratory quotients of leaves containing, 26–7
apple fruit, alcohol and aldehyde in, 78–80
— —, anaerobic respiration of, 71–3, 75, 131–6
— —, Pasteur effect in, 130–6
— —, respiration of, 38–40, 50–51, 71–3, 75, 131–6
— —, terminal oxidases in, 127–128
— seed, effect of temperature on respiratory quotient of, 27
Appleman, C. O., 76
artichoke tubers, effect of oxygen concentration on respiration of, 49–50
— —, Pasteur effect in, 76, 131
Artocarpus integrifolia, respiration of, 42
Arum italicum, evolution of heat by flowers of, 3, 139
— *maculatum*, anaerobic respiration of, 69
— *sp.*, rapid respiration in spadix of, 129–30
ascorbic acid oxidase, 113, 116–117, 125, 128–30
asparagine, 21–4, 91

161

Asparagus, effect of water content on respiration of, 40
—, protein as respiratory substrate in, 24
—, respiratory quotient of, 24
aspartic acid, 21, 91
Aspergillus niger, anaerobic respiration of, 64, 88
— —, glucose oxidase in, 88, 117
— —, respiration intensity of, 32–3
— —, respiratory quotients of, on different media, 15–16
Aspidistra, respiratory quotient of leaves of, 17
ATP, *see* adenosine triphosphate
Aubert, E., 26, 33
Aucuba, respiratory quotient of leaves of, 17
Audus, L. J., 56
avocado, respiration of, 38
azide as inhibitor of enzyme actions, 115–18, 128
— as inhibitor of respiration, 68, 128

Bach, A., 113
Bacillus denitrificans, 6
— *hydrogenes*, 7
— *mesentericus vulgatus*, respiration intensity of, 33
— *pantotropus*, 7
— *phosphorescens*, 6
Bacteria, respiration of, 6–7, 32–33
Baker, D., 126
banana, respiration of, 38
Barcroft, J., 11
Barker, J., 54, 57
barley, acetaldehyde in, 87
—, anaerobic respiration of, 59–60, 64–5
—, asparagine in starved leaves of, 23–4
—, carboxylase in, 110
—, pyruvic acid in, 110
—, respiration in phosphorus-deficient, 109
—, respiratory quotient of starved leaves of, 23
—, terminal oxidase in, 128

Bartholomew, E. T., 24
bean, broad, *see Vicia Faba*
Beck, L. V., 105
beetroot, effect of organic acids on respiration of, 136
—, effect of salts on respiration of, 54
—, effect of wounding on respiration of, 57
—, Pasteur effect in, 76, 131
—, respiratory quotient of, 24, 136
Beevers, H., 129
Beggiatoa, 7
Begonia, respiratory quotient of leaves of, 17
Bellamy, F., 60
Bennet-Clark, T. A., 54, 136–7
Bennett, E., 111
Bennett, J. P., 24
benzene, effect of, on respiration, 68
Bexon, D., 54, 136–7
Biale, J. B., 38–9
blackberry, anaerobic respiration of, 73
—, isocitric acid in, 123
Blackman, F. F., 23, 38–40, 42, 46, 50, 71–3, 131–5, 137
Böhm, J., 57
Bonnier, G., 10–11, 18, 21–2, 41
Bonner, J., 110–11, 123, 128
Boswell, J. G., 77, 125–7, 129
Bramley's Seedling apple, extinction point for, 79
— — apple, respiration of, 38–9, 71–3, 131–5
Brassica napus, anaerobic respiration of, 69
Briggs, G. E., 34, 37–8
broad bean, *see Vicia Faba*
Brown, A. H., 128
Brown, R. G., 76
Bryophyllum, isocitric acid in, 123
Buchner, E., 98
buckwheat, *see Fagopyrum*
bulb of *Narcissus poeticus*, respiratory quotient of, 29
bulky organs, respiration in, 26, 77–9
Bunting, A. H., 110
Burström, H., 54–5
butyric bacteria, 6

cabbage, ascorbic acid oxidase in, 116
Cactaceae, respiration of, 26
Cannabis sativa, anaerobic respiration of, 69
— —, chemical changes during germination of seeds of, 20–1
Cannizzaro reaction, 100, 108
Cantharellus cibatus, anaerobic respiration as, 69
carbohydrate as respiratory substrate, 9, 14–20, 43, 64, 90, 130–1, 133–4
carbon dioxide concentration, effect of, on respiration, 28, 51–3
— — zymasis, 79
carbon monoxide as inhibitor of enzyme actions, 115–18, 126–8
— — as inhibitor of respiration, 126–8
carbonic anhydrase, 112, 118
carboxylase, 91, 100, 107, 110–111, 118, 120, 124, 130
carpels, respiration intensity of, 33–4
carrot root, anaerobic respiration of, 73
— — effect of inhibitors on aerobic and anaerobic respiration of, 68, 89, 127
— — effect of oxygen concentration on respiration intensity of, 50
— —, effect of oxygen concentration on respiratory quotient of, 28–9
— —, effect of salts on respiration intensity of, 54–5
— —, effect of wounding on respiration of, 57
— —, internal atmosphere of, 78
— —, Pasteur effect in, 76, 131
— —, terminal oxidases in, 127–128
catalase, 112, 117–18
catechol oxidase (catecholase), 113–16, 125–30
Cereus macrogonus, respiration intensity of, 33

cherry fruit, anaerobic respiration of, 73
cherry laurel leaves, effect of carbon dioxide concentration on respiration of, 51
— — leaves, effect of chloroform on respiration of, 55–6
— — leaves, effect of mechanical stimulation on respiration of, 56–7
— — leaves, effect of temperature on respiration of, 46
— — leaves, respiratory quotient of, 17
chloroform, effect of, on respiration, 55–6
Chodat, R., 113
Christian, W., 104
Chrysanthemum, respiratory quotient of leaves of, 17
Chudiakov, N. V., 59
cis-aconitic acid, 119–20
citric acid (and citrate), 119–20, 123, 126
Citrus fruits, climacteric rise in, 39
Clausen, H., 46
climacteric rise, 38–40
cocarboxylase, 107, 121
coenzyme 1, 98, 105, 108–9, 112, 120, 122
coenzyme 2, 112, 120
coenzyme A, 119
Cohen, Clara, 83
combustion and respiration, 13–15
copper enzymes, 115, 117
Cori ester, 103
cotton, citric acid in, 123
Coulthard, C. E., 117
course of aerobic respiration, 109–38
— of anaerobic respiration and fermentation, 97–109
cozymase, *see* coenzyme 1
Crassulaceae, respiration of, 26
Crenothrix polyspora, 7
Cruickshank, W., 59
Cucurbita, anaerobic respiration of, 64, 69, 74–5
—, effect of wounding on respiration of, 57

Cucurbita fruits, internal atmosphere of, 78
cyanide as inhibitor of enzyme actions, 115–18, 126–7
— as inhibitor of respiration, 55, 68, 126–7
cytochrome, 112, 115–16, 121
cytochrome oxidase, 113, 115–116, 118, 121–2, 125–30
Czerny, F., 83

Damodaran, M., 123
De Boer, S. R., 16
decarboxylase, *see* carboxylase
decarboxylation, 100, 107, 110–111, 119–20, 124, 137
dehydrogenases, 112–13, 118–21
Demoussy, E., 16–17
Denny, F. E., 49, 53, 78
Dent, K. W., 24, 64, 74, 76, 136
desiccation, effect of, on respiration, 40–1
Detmer, W., 20–1
Devaux, H., 78
development, respiration intensity during, 34–43
Diakanov, N. W., 84–5
diaphorase, 122
dihydroxyacetone, 99
— phosphate, 104–5
diphosphoglyceric acid, 105–6, 108, 140
diphosphopyridine nucleotide, *see* coenzyme 1
disaccharides, utilization of, in respiration, 90
dissimilation, 1
'diurnal respiration', 4
Dixon, M., 106–7
DPN, *see* coenzyme 1

Eaton, F. M., 123
edible pea, *see* Pisum sativum
Elodea, absence of anaerobic respiration in, 86
Embden, G., 146
energy, release of, in respiration, 1, 5–9, 13–15, 32, 66, 138–142
—, utilization of, in respiration, 8, 137–8, 141–2

energy-rich phosphate bonds, 102–3, 106, 139–41
— transfer in respiration, 138–42
enolase, 107
environmental factors and respiration, 43–56
enzymes, oxidizing, 112–30
— concerned in fermentation and respiration, 15, 81–3, 90, 98–101, 103–31, 133, 140
Ergle, D. R., 123
ether, effect of, on respiration, 68
ethyl alcohol, effect of, on respiration, 68
— —, production of, in anaerobic respiration, 60–6, 76, 78
— — in fruits, 78–80
Euler, H., 98
exchanges of gases in respiration, 1–5, 9–32
external factors in respiration, 43–58, 66–8
extinction point, 77

factors, external, in respiration, 43–58
—, internal, in respiration, 40–3
facultative anaerobes, 6
Fagopyrum, anaerobic respiration of germinating seeds of, 74–5
fat seeds, respiratory quotient during germination of, 17–20, 26
— —, respiratory quotient during maturation of, 25–6
fats as substrate for respiration, 16–21, 32, 90–1
Fenn, W. O., 11
fermentation, alcoholic, 5–6, 60–61, 97–109, 122, 140–1
— (anaerobic respiration), 5–6, 59–80, 82–9, 108–9, 122, 140–1
Fernandes, D. S., 35–6, 44–5, 66–8
ferrous iron, utilization of, in respiration, 7
Fidler, J. C., 79, 80, 131
flavin, 112, 122
flax, *see* Linum usitatissimum
'floating respiration', 23, 42, 51

INDEX

flowers, respiration of, 3, 33–4, 37, 129–30
Fontinalis, respiration of, 55
food reserve in seeds, 17, 21
fructose, constitution of, 94–6
— as respiratory substrate, 40, 90
fructose-1,6-phosphate, 97, 99, 101, 104, 110–11, 140
fructose-6-phosphate, 103–4, 110, 140
fruits, alcohol and aldehyde in, 78–80
—, internal atmosphere of, 78
—, Pasteur effect in, 130–7
—, zymasis in, 78–80
fumaric acid, 119, 121–2, 141
— dehydrogenase (fumarase), 119, 121, 124
fungi, anaerobic respiration of, 63–4, 83
—, respiration of, on different media, 15–16, 32–3
—, respiratory quotients of, 15–16
furan, 94
furanose sugars, 94–6

Garreau, 4
gaseous exchange, 1–5, 9–32
Geddes, W. F., 139
geotropic curvature and respiration, 57
Gerber, C., 25, 78
Gerhart, A. R., 47
germination, oxygen necessary for, 2–3
—, respiration intensity during, 41
—, respiratory quotient during, 17–20
germinating seeds, anaerobic respiration of, 74–6
Gladiolus corms, internal atmosphere of, 53, 78
— —, respiration of, 53
glucose, constitution of, 92–6
— as respiratory substrate, 90
— oxidase, 88, 117
glucose-1-phosphate, 90, 96–7, 103, 110
glucose-6-phosphate, 96–7, 103, 140

glyceraldehyde, 99–100
glyceraldehyde-3-phosphate, 104
glycerol production during fermentation, 100–1
glycolysis, 99–101, 111, 118, 124, 130–6, 141
glycosides, utilization of, in respiration, 90
Goddard, D. R., 28–9, 50, 68, 76, 111, 127–8
Godlewski, E., 62
Gottschalk, A., 83
Gramineae, food reserves in seeds of, 17
grand period of respiration, 36, 46
grape, anaerobic respiration of, 70, 73
—, wild, respiratory quotient of leaves of, 17
Griffiths, D. G., 39
growth rate and respiration, 36
guaiaconic acid, 113
guaiacum gum, 113–14
guava, Pasteur effect in, 131
Gustafson, F. G., 38, 80
Guthrie, J. D., 123
gynaecium, respiration intensity of, 33–4

Hackney, F. M. V., 127, 129
Hanes, C. S., 54, 103–4, 110
Harden, A., 97–9
haricot, respiratory quotient of leaves of, 17
Harrington, G. T., 27
Hartree, E. F., 117
Haworth, W. N., 92
Heard, C. R. C., 110
heat, evolution of, in respiration, 3, 8, 14, 138–9
Helianthus annuus, anaerobic respiration of, 64, 69, 74–5
— —, respiration intensity of, during development, 34
hematin, 113, 115
hemicelluloses, utilization of, in respiration, 90
hemp, chemical changes during germination of seeds of, 20–1
Heracleum giganteum, anaerobic respiration of, 69
heterohexoses, 134

hexokinase, 103
hexose as respiratory substrate, 15–16, 90–1
—, diphosphate, *see* fructose-1,6-phosphate
— monophosphates, 99, 110
hexuronic acid, 116
Hill, G. R., 73
Hillhousia, 7
Hopkins, E. F., 58
horse-radish, peroxidases of, 113
Hübbenet, E., 83
Hydnum repandum, anaerobic respiration of, 69
Hydrangea otaksa, effect of carbon dioxide on respiration of leaves of, 51–2
hydrocyanic acid, *see* cyanide
— — zymasis, 79–80
hydrogen, utilization of, in respiration, 7
— bacteria, 7
— sulphide, utilization of, in respiration, 7
Hydrogenomonas, 7
hydroxylamine as inhibitor of catalase, 118

Inamdar, R. S., 42
indophenol oxidase, 113
Ingen-Housz, J., 3
inhibitors of enzyme actions, 88–89, 110, 115–18, 125–6
— of respiration, 55–6, 68, 125–126
inorganic salts, effect of, on respiration, 54–5
intensity of respiration, 30–58
intermediates in respiration, 64–65, 86–7, 99–101, 103–12, 118–24
internal factors in respiration, 40–3
'intramolecular respiration', 60
inulase, 90
inulin, utilization of, in respiration, 90
iodoacetic acid (and iodoacetate) as inhibitor of respiration, 89
— — (and iodoacetate) as inhibitor of zymase, 88–9, 110, 118

ionized air, effect of, on respiration, 53
iron bacteria, 7
— catalysis system, 55
— enzymes, 113, 116, 118
Irving, A. A., 56
isocitric acid, 119–20, 123
— dehydrogenase, 119–20, 123
Ivanov, L., 83
Ivanov, N., 83
ivy, respiratory quotient of leaves of, 17

James, G. M., 110
James, W. O., 87, 109–11, 117–118, 124, 128–9
Jensen, P. Boysen, 62–3, 70, 73, 88–9
Jerusalem artichoke tubers, effect of oxygen concentration on respiration of, 49–50
Jönsson, B., 41

Karczag, L., 99
Karslen, A., 68
katharometer, 11
Keilin, D., 115–17
Kermack, W. O., 106
α-keto-β-carboxyglutaric acid, 119–20
α-ketoglutaric acid, 119–20, 123
— dehydrogenase, 119–20
ketoses, constitution of, 94–5
Kidd, F., 34, 37–8, 40, 48, 51–2
Kiesling, W., 107
Klein, G., 87
Kobel, M., 110
Kosinski, I. 54
Kostychev, S. (Kostytschew, S.), 32, 60, 62–4, 83, 85–7, 89, 118
Kraus, G., 139
Krebs cycle, 119, 123–4, 129–30, 137, 141
Kuijper, J., 47

laccase, 114
Lactarius piperatus, anaerobic respiration of, 69
lactic acid production in anaerobic respiration, 65, 99

lactic bacteria, 6
Laine, I., 123
Lamarck, J. B., 3, 139
Lathyrus odoratus, respiration of, during germination, 36, 38, 74–5
Lavatera olbia, respiratory intensity of various organs of, 34
Lavoisier, A. L., 2–3
Leach, W., 64, 74, 136, 139
Leavenworth, C. S., 23
leaves, effect of wounding on respiration of, 57
—, 'floating' and 'protoplasmic' respiration of, 23
—, respiration intensity of, 34, 37
—, respiratory quotients of, 16–17, 23, 26–7
Lechartier, G., 60
Leonard, E. R., 38
Levy, H., 127
lichens, resistance of, to desiccation, 41
Liebig, J. v., 4
light, effect of, on respiration, 48
Ligustrum vulgare, anaerobic respiration of, 69
lilac, see *Syringa*
lily, respiratory quotient of leaves of, 17
linseed, see *Linum usitatissimum*
Linum usitatissimum, respiratory quotient during germination of seeds of, 18–20, 22, 26
— —, respiratory quotient during maturation of seeds of, 25–6
lipase, 90
Lipmann, F., 102
Livingston, E., 51
Lohmann, K., 102–4
Lundegårdh, H., 54–5
Lundsgaard, E., 88
Lupinus albus, respiration intensity of, 33
— *luteus*, anaerobic respiration of, 69
— —, chemical changes in seeds of, during germination, 21
— —, respiratory quotient of seedlings of, 21–2
Lupinus, sp. seedlings, effect of temperature on respiration of, 46
Lyon, C. J., 83, 86

McCall, E. R., 123
Machlis, L., 128
Magness, J. R., 77
Mahonia, respiratory quotient of leaves of, 17
Maige, A., 54
Maige, G., 34
maize, see *Zea mais*
malic acid (and malate), 26, 119, 121, 123–4, 136–7, 141
— — dehydrogenase, 119, 121, 124
Malpighi, M., 2
maltase, 90
Mamillaria cliphatidens, respiration intensity of, 33
manganese catalysis system, 55
Mangin, L., 10–11, 18, 21–2, 41
mango, Pasteur effect in, 131
mangold, Pasteur effect in, 131
manometric methods, 10–11
Maquenne, L., 16–17, 86
Marsh, P. B., 28–9, 50, 68, 76, 127
Matthaei, G. L. C., 46
maturation of linseed, respiratory quotient during, 25–6
Mayer, A., 35
measurement of respiration, 9–12
mechanical stimulation, effect of, on respiration 56–7
mechanism of respiration, 81–142
Meeuse, B. J. D., 111
meristems, respiration intensity of, 33
Merry, J., 128
Mesembryanthemum, respiration of, 26
— *deltoides*, respiration intensity of, 33
methyl glyoxal, 99–100
Meyerhof, O., 101, 103–6, 133, 136
micro-eudiometers, 10
Mnium undulatum, effect of water content on respiration of, 41
Mohl, H. v., 4

monoiodoacetic acid, *see* iodoacetic acid
monophenol oxidase, 115
monophosphoglyceric acid, 106
Morgan, E. J., 111
mosses, effect of water content on respiration of, 41
Müller, D., 88, 117
Müller-Thurgau, H., 47
mustard, white, effect of carbon dioxide concentration on respiration of seedlings of, 51–2
Myrbäck, K., 98

Nabokich, A. J., 62
Nägeli, C., 84
Nair, K. R., 123
Narcissus poeticus, respiratory quotient of bulb of, 29
Needham, D., 106
Nelson, J. M., 126
Neuberg, C., 83, 99–101, 107, 110
Newton Wonder apple, extinction point for, 79
Nicolas, M. G., 26–7, 54
nitrating bacteria, 6
nitrifying bacteria, 6
nitrites, utilization of, in respiration, 6
Nitrobacter, 6
4-nitrocatechol as inhibitor of catechol oxidase, 126
Nitrococcus, 6
Nitrosomonas, 6
'nocturnal respiration', 4
Norval, I. F., 87, 110
notatin, 117

oat coleoptile, effect of pyruvate on respiration of, 111
— —, hexose phosphates in, 110
— —, terminal oxidase in, 128
obligate anaerobes, 6
Olney, A. J., 38
onion, pyruvic acid in, 111
Onoclea sensibilis, carbonic anhydrase in, 118
optimum temperature for respiration, 44–5

Opuntia, respiratory quotient of, 26
orange, Pasteur effect in, 130
organic acids in anthocyanin-containing leaves, 27
— — in respiration, 26–8, 48, 65, 131, 136
— — in succulent plants, 26
— substances, effect of, on respiration, 55–6
'organization resistance', 39–40
Orobanche ramosa, anaerobic respiration of, 69
oxalacetic acid, 119, 121–3, 137–138
oxalic acid in *Mesembryanthemum*, 26
oxidases, 112–17, 122–30
oxidation processes in respiration, 5–7, 9, 13–14, 16, 20–23, 27, 111–12, 118–38, 140–1
oxidative anabolism, 130–8
oxidizing enzymes, 112–30
oxygen concentration, effect of, on respiration intensity, 48–51, 76–7, 127, 133
— —, effect of, on respiratory quotient, 28–9, 76
oxygen necessary for seed germination, 2–3
oxygen respiration, *see* aerobic respiration
oxygenase, 114

paeony, effect of water content on respiration of, 40
Palladin, V. I., 1, 4, 12, 54, 62, 86
Papaver, respiratory quotient of leaves of, 17
— *rhoeas*, respiration intensity of various organs of, 34
Parija, P., 38–9, 71, 134
parsnip roots, Pasteur effect in, 76
Pasteur, L., 60, 76
Pasteur effect, 76, 130–8
pea, edible, *see Pisum sativum*
peach, respiration of fruit of, 38
pear, respiration of fruit of, 38
—, respiratory quotient of leaves of, 17

INDEX

Penicillium glaucum, aerobic and anaerobic respiration of, 88
— *notatum*, glucose oxidase in, 117
— *resticulosum*, glucose oxidase in, 117
pentoses, constitution of, 94–5
Pentstemon gentianoides, respiration intensity of various organs of, 34
peptone as substrate in anaerobic respiration, 64
periodicity, seasonal, relation of respiration to, 41–3
peroxidases, 112–14
petals, respiration intensity of, 33–4
Pettenkofer tubes, 12
Pfeffer, W., 1, 4, 8, 60, 69–70, 84–5, 87, 89, 118
Pflüger, E., 60, 84
Phillips, J. W., 64–5
phosphate bonds, energy-rich, 102–3, 106, 139–41
phosphoenolpyruvic acid, 107
phosphoglucomutase, 103
phosphoglyceric acid (and phosphoglycerate), 101, 105–6, 110–11, 118, 140
phosphoglyceric aldehyde, 104–106, 140–1
— phosphokinase, 106
phosphohexose isomerase, 103
phosphokinases, 106–7
phosphopyruvic acid, 107, 119
phosphoric acid (and phosphate), part played in fermentation and respiration by, 83, 96–99, 101–11, 140–1
phosphorylase, 90, 103
phosphorylated hexose, 134
phosphorylation, 99, 101, 103, 106, 140–1
phosphotriose isomerase, 105
Photinia glabra, respiratory quotients of green and red leaves of, 27
Phycomyces, respiratory quotients of, 16
Pirschle, K., 87
Pisum sativum, anaerobic respiration of germinating seeds of, 62, 66–8, 83

Pisum sativum, effect of carbon dioxide concentration on respiration of, 52–3
— —, effect of temperature on respiration of seedlings of, 44–6
— —, fructose-1,6-diphosphate from, 104, 110
— —, α-ketoglutaric acid in, 123
— —, oxalacetic acid in, 123
— —, phosphorylase in, 103
— —, respiratory intensity of seeds and seedlings of, 35–6
— —, respiratory quotient of leaves of, 17
— —, respiratory quotient of seedlings of, 29
plasmolysis and respiration, 54
Platenius, H., 24
poisons, effect of, on respiration, 55–6, 68
pollination and respiration, 57
polyphenol oxidase, 113–14
polysaccharides, utilization of, in respiration, 90
Polzeniusz, F., 62
poplar flowers, acetaldehyde in, 83
poppy, *see Papaver*
potato tuber, anaerobic respiration of, 62–3, 130
— —, citric acid in, 123
— —, effect of catechol on respiration of, 125–6
— —, effect of handling on respiration of, 56–7
— —, effect of oxygen concentration on sprouting of, 53
— —, effect of salts on respiration of, 54
— —, effect of sugar concentration on respiration of, 54
— —, effect of water content on respiration of, 41
— —, effect of wounding on respiration of, 57–8
— —, internal atmosphere of, 77
— —, phosphorylase in, 103
— —, respiratory quotient of, 24
— —, terminal oxidases in, 125–128

Potter, N. A., 39
Pratt, H. K., 38
Preston, C., 54
Prianischnikov, D., 21
Priestley, J., 2
privet, respiratory quotient of leaves of, 17
protease, 91
proteins, oxidation products of, 21–2
— as respiratory substrate, 21–4
'protoplasmic respiration', 23, 42
Prunus cerasifolia, respiratory quotients of green and red leaves of, 27
Pucher, G. W., 23, 123
Purievitch (Puriewitsch), K., 15–16
pyran, 93
pyranose sugars, 93–6, 133–4
pyruvic acid in barley, 110–11
— — in fermentation and respiration, 64, 87, 99–101, 107–8, 110–11, 119, 122–124, 130, 137–8
— — in onion, 111

quinic acid as substrate for *Aspergillus*, 32–3
quotient, respiratory, 15–30, 49, 75–6, 136

radioactive substances, effect of on respiration, 53
Ramstad, P. E., 139
Raphiolepsis ovata, respiratory quotients of green and red leaves of, 27
'reserve celluloses', utilization of, in respiration, 90
reserve food in seeds, 17, 21
resorcinol as inhibitor of catechol oxidase and respiration, 127
respiration intensity, 30–58
respiratory quotient, 15–30, 49, 75–6, 136
— substrate, 5–7, 9, 14–24, 30, 32, 39–40, 43, 81–2, 84–6, 89–97, 130–1, 141
rhubarb, respiratory quotient of leaves of, 17

Rhus spp., laccase in, 114
rice seedlings, effect of oxygen concentration on respiration of, 51
— —, terminal oxidases in, 128
Richards, F. J., 109
Richards, H. M., 57
Ricinus, anaerobic respiration of, 64, 74–5
—, respiratory quotient of leaves of, 17
ripening of linseed, respiratory quotient during, 25–6
Rischavi, L., 35
Robertson, R. N., 54–5
rose, respiratory quotient of leaves of, 17
Roux, E. R., 38

Sachs, J. v., 1–2, 4, 60, 84
salt respiration, 55
salts, effect of, on respiration, 54–5
Saussure, T. de, 3, 8, 35, 51
Schade, A. L., 126–7
Scheele, C. W., 2
Scheloumow, A., 83
Schneider-Orelli, O., 47
seasonal periodicity, relation of respiration to, 40–3
Sedum dendroideum, respiration intensity of, 33
seedlings, course of respiration during development of, 34–8
—, effect of temperature on respiration of, 44–6
seeds, food reserves of, 17, 21
—, germinating, anaerobic respiration of, 74–6
—, oxygen necessary for germination of, 2–3
—, respiration intensity during germination of, 34–8, 41
—, respiration intensity during maturation of, 41
—, respiratory quotient during germination of, 17–20, 26
—, respiratory quotient during maturation of, 25–6
Seifriz, W., 61
senescence, respiration during, 38–40, 48

sepals, respiration intensity of, 33-4
shade plants, respiration intensity of, 33
Sinapis alba, anaerobic respiration of, 69-70, 88
Singh, B. N., 42
Snow, D., 139
sorrel, respiratory quotient of leaves of, 17
soya bean, heating of, in storage, 139
spinach, protein as respiratory substrate in, 24
spindle tree, respiratory quotient of leaves of, 17
Spirophyllum ferrugineum, 7
Spoehr, H. A., 12
stages in aerobic respiration, 109-38
— in fermentation and anaerobic respiration, 97-109
stamens, respiration intensity of, 33-4
starch, utilization of, in fermentation and respiration, 90, 103
starved leaves, respiration of, 25, 54
Stenlid, G., 128
Steward, F. C., 54
Stich, C., 28-9, 49, 57
Stiles, W., 24, 76
stimulation, effect of, on respiration, 56-7
Stoklasa, J., 83
strawberry fruits, effect of temperature on respiration of, 47
streaming of protoplasm, 8
substrate for respiration, 5-7, 9, 14-24, 30, 32, 39-40, 43, 81-2, 84-6, 89-97, 130-1, 133-4, 141
succinic acid (and succinate), 119, 123, 136, 141
— dehydrogenase, 119, 121, 123-4
succulent plants, respiration of, 26, 33, 77
sucrase, 90
sucrose, utilization of, in respiration, 90
sugar as respiratory substrate, 14-17, 32, 40, 84-6, 90

Sugar concentration, effect of, on respiration, 53-4
sugars, constitution of, 92-6
sulphide as inhibitor of enzyme actions, 115-18
sulphur bacteria, 7
sunflower, respiration intensity of, during development, 34, 37
swede, terminal oxidase in, 129
sweet pea, see *Lathyrus odoratus*
Syringa flowers, effect of temperature on respiration of, 46
Szent-Györgyi, A., 116

Tankó, B., 104, 110
Taylor, D. L., 51, 128
temperature, effect of, on anaerobic respiration, 66-8
—, effect of, on respiration intensity, 38-9, 43-8, 66-8
—, effect of, on respiratory quotient, 27-8
temperature coefficient of respiration, 46-7, 67-8
terminal oxidases, 123, 125-30
testa, influence of, on respiration, 77
Theorell, H., 113
thermophile bacteria, 6
thiamin, 107
Thiobacillus, 7
thiosulphate bacteria, 7
Thiothrix, 7
Thomas, M., 6, 62, 77-80
Thornton, N. C., 53, 78
Thunberg, T., 11
time factor, 44-6
TPN, see coenzyme 2
tobacco, respiratory quotient of leaves of, 17
—, utilization of amino-acids in starved leaves of, 23
tomato, Pasteur effect in, 131
—, production of alcohol in, 80
—, respiration of, 38
Trifolium pratense, carbonic anhydrase in, 118
triosephosphate, 104-5
— dehydrogenase, 105, 112, 118
— isomerase, 104-5
triosis, 99

triphosphopyridine nucleotide, *see* coenzyme 2

Triticum sativum (*T. vulgare*), anaerobic respiration of, 69

— — (*T. vulgare*), effect of oxygen concentration on respiration of seedlings of, 49, 51

— — (*T. vulgare*), effect of poisons on respiration of, 68

— — (*T. vulgare*), effect of salts on respiration of seedlings of, 54

— — (*T. vulgare*), effect of temperature on respiration of, 46

— — (*T. vulgare*), respiration intensity of, 33, 36

— — (*T. vulgare*), respiratory quotient during germination of seeds of, 18–20

— — (*T. vulgare*), respiratory quotient of leaves of, 17

— — (*T. vulgare*), respiratory quotient of seedlings of, 29

— — (*T. vulgare*), terminal oxidase in embryo of, 128

Tropaeolum majus, anaerobic respiration of leaves of, 70, 88

Turner, J. S., 76, 89, 99

turnip, respiratory quotient of leaves of, 17

tyrosinase, 114–15

Van't Hoff rule, 46

Vennesland, B., 124, 137

Vicia Faba, anaerobic respiration of, 68–9

— —, effect of wounding on respiration of, 57

— —, respiration intensity of, 33, 54

Vickery, H. B., 23, 123

Vignol, M., 33

vine, respiratory quotient of leaves of, 17

Virtanen, A. I., 123

vitamin B_1, 107

— C, 116

Wakeman, A. J., 23

Warburg, O., 11, 104

Wardlaw, C. W., 38

water content, relation of respiration to, 40–1

Wehner, O., 55

West, C., 34, 37–9, 48

wheat, *see Triticum sativum*

Wheldale, M., 113

Whiting, G. C., 77, 125–6

Wortmann, J., 68

wounding, effect of, on respiration, 57–8

Wildman, S. G., 123

Wright, H. C., 139

xanthine oxidase, 117

yeast, anaerobic respiration of, 69

— fermentation, 5, 60–1, 82–4, 87, 97–109

Yemm, E. W., 23, 91

Young, R. E., 39

Young, W. J., 97–9

Zaleski, W., 83

Zea mais, anaerobic respiration of germinating seeds of, 74–5

— —, respiratory quotient of leaves of, 17

— —, respiratory quotient of seedlings of, 29

zymase, 62, 82–3, 86–8, 91, 98, 133

zymasis, 6, 62, 79–80

zymohexase, 104